The SEAFOOD CAPITAL of the World

The SEAFOOD CAPITAL of the World

BILOXI'S MARITIME HISTORY

Edmond Boudreaux

THE History PRESS

Published by The History Press
Charleston, SC 29403
www.historypress.net

All images by the author unless otherwise noted.

First published 2011

Manufactured in the United States

ISBN 978.1.60949.284.7

Library of Congress Cataloging-in-Publication Data
Boudreaux, Edmond.
The seafood capital of the world : Biloxi's maritime history / Edmond Boudreaux.
p. cm.
Includes bibliographical references and index.
ISBN 978-1-60949-284-7
1. Seafood industry--Mississippi--Biloxi--History. 2. Fish trade--Mississippi--Biloxi--History. 3. Fishing boats--Mississippi--Biloxi--History. 4. Ethnology--Mississippi--Biloxi--History. 5. Biloxi (Miss.)--Economic conditions. 6. Biloxi (Miss.)--Commerce. 7. Biloxi (Miss.)--History, Naval. 8. Biloxi (Miss.)--Social life and customs. I. Title.
HD9458.B55B68 2011
338.3'720976213--dc23
2011019692

Contents

Acknowledgements

While many years of research have gone into this book, certain people and organizations have made this book possible. In 1985, I joined the Mississippi Coast Historical & Genealogical Society. I was mentored by a special group of local historians that included Julia Cook Guice, Mary Louise Adkinson, Murella Hebert Powell and Kat Bergeron. The society was very active in preservation of historical documents, pictures, etc. This group fought hard to preserve our Mississippi Gulf Coast history and still does so today.

Another organization I joined in 1985 was the Gulf Coast Chapter of the Mississippi Archaeology Association. This group of amateurs and professionals worked together to preserve, record and protect archaeological sites along the Mississippi Gulf Coast. This group is instrumental in educational programs and providing volunteers for archaeological excavations along our Mississippi Gulf Coast.

Also in 1985, I became involved with the Maritime & Seafood Industry Museum located in Biloxi. Here I was mentored by Dr. Val Husley, the curator of the museum. In 1985, the museum was in the process of being created to preserve the maritime and seafood industry history of Biloxi and the Mississippi Gulf Coast. It developed into a first-class museum, but in 2005, Hurricane Katrina destroyed the building and all the exhibits less than one year after a total remodeling and addition to the museum. Today, it has a temporary location in the Edgewater Mall and is in the process of rebuilding. The museum's records, documents and pictures played a big part in the research of this book.

The Local History and Genealogy Department of the Harrison County Library System also deserves special thanks. Its collection of historic photographs, documents, newspapers, directories and other records has been instrumental in the research of this book.

Thanks to the Mississippi Department of Archives and History, which worked closely with me on archaeological sites and archaeological investigations along the Mississippi Gulf Coast. Thanks to Samuel O. McGahey, my early mentor, and Patricia Galloway. Special thanks to Jack Elliott, a friend and mentor in archaeological research and methods.

I would also like to thank The History Press for recognizing the history of Biloxi and the Mississippi Gulf Coast through the printing of this book.

To Virginia, my wife, and my children, Tony, Brandon and Marcus, a very special thanks for editing my work and keeping me straight.

INTRODUCTION

To understand the history of Biloxi, one has to understand its maritime and seafood industry history. Biloxi is a peninsula, with the waters of the Mississippi Sound and Biloxi Bay surrounding it on three sides. Its historical or written history is over three hundred years old, but its ancient history is over ten thousand years old.

The early Native Americans realized that the peninsula of Biloxi would provide them with an abundance of shellfish, fish, crustaceans and other fresh and saltwater sea life. Also in abundance were deer, rabbits, birds and other small mammals. Their shallow-drafted dugout canoes were ideal for the rivers, bays and shallow water of the Mississippi Sound.

Biloxi's colonial maritime history was tied to the French, English and Spanish. Beginning with the arrival of the French in February 1699, Biloxi's location, timber and backwaters would attract attention. During the colonial periods, Biloxi was a perfect place for timber and other raw materials needed for the construction of sailing vessels. Also, the abundance of seafood and other animals made it attractive to the early settlers.

After Mississippi became part of the United States in 1811, commerce and tourism began to develop. Sailing vessels plied the Mississippi Sound between New Orleans and Mobile. Biloxi soon became one of the most visited ports in this area. Its raw materials and seafood were harvested and taken to market by sailing and steam vessels. It would slowly develop into a summer resort for the elite of New Orleans and rich plantation owners. Hotels and boardinghouses began to dot the Biloxi beaches and downtown district.

In 1848, the United States government built a lighthouse at Biloxi for the safety of mariners and their passengers. The arrival of the Civil War in 1861 marked the beginning of a long struggle. The war would have an adverse effect on Biloxi's economy that would last into the 1870s. In 1871, the railroad was completed between New Orleans and Mobile. One by one, railroad stations began to dot the Mississippi Gulf Coast, and soon the rails were full of tourists. Visitors began to arrive from northern cities in the winter and southern cities in the summer. Additional hotels and boardinghouses began to be constructed in Biloxi.

In 1881, Lopez, Elmer and Company, the first seafood-canning factory, opened in Biloxi. These factories would attract a new breed of Biloxians. The first to arrive were Slavonians from Austria-Hungary. With more seafood workers needed, the Bohemians would fill the void. These Bohemians were seasonal workers who arrived by rail from Baltimore, worked for several months and then returned to Baltimore. In due time, this source of transient workers ended, and the Cajuns from South Louisiana were recruited. The Slavonians, Bohemians and Cajuns moved into the Biloxi canning district, building homes and creating a new life. Due to their religious beliefs and their strong connection to the Mississippi Sound, an annual Blessing of the Fleet was created in 1929.

Hurricanes have devastated the Mississippi Gulf Coast throughout its history. Some of the most notable occurred in 1722, 1893, 1915 and 1947. In recent times, Hurricane Camille in 1969 and Hurricane Katrina in 2005 have dealt Biloxi catastrophic blows. Like past generations, current Biloxians immediately started rebuilding and improving their city.

In 1992, a new industry arrived in Biloxi when the State of Mississippi legalized dockside gaming. The success of gaming in Biloxi would improve tourism, seafood sales and other aspects of life in the city. Today, Biloxi has strong ties to the Gulf and Biloxi Bay's waters. The Gulf and coastal backwaters are still a source of food, revenue and personal pleasure for many Biloxians. Biloxi has strong ties to its maritime past with eyes on its development and future.

Chapter 1

Coastal Native Americans

When considering the maritime history of Biloxi, we must include the Native Americans who were the first to enjoy the bountiful sea life of the rivers, bayou, bays and the Mississippi Sound. These were the first individuals to make a living from the abundant seafood that surrounded the Biloxi Peninsula. Seafood from the Mississippi Sound in the form of oysters, mussels, crabs, shrimp, fish and other sea life was first harvested by Native Americans. The Gulf Coast, with all its natural resources, became a much-visited and inhabited area. Significant Native American cultural developments are divided into four periods: Paleoindian, Archaic, Woodland and Mississippian (according to Blitz and Mann's *Fisherfolk, Framers, and Frenchmen*).

Along the Gulf Coast, Paleoindian visits have been proven by the recovery of stone spear points and tools used during this period. Due to the colder climate, sea level along the Gulf Coast was also lower, and many Paleoindian sites are now under Gulf waters. The Paleoindians roamed in small family bands along the shores, inlets and rivers. They lived off the land, hunted for meat and used animal skins for clothing and shelter. In coastal areas, fish, crustaceans, shellfish, birds, small animals and various plant life were probably a supplement to their diet. The Paleoindian Period ended about 8000 BC, when the climate began to warm and the larger mammals, who were on the brink of extinction, began migrating northward with the retreating glaciers.

As the climate began to warm, the sea level submerged former coastlines. These changes forced the hunters to seek smaller game such as white-tailed

Local collection of Mississippi Gulf Coast Native American points.

deer, turkeys, wildcats, bears, rabbits and other small forest animals and, from the streams, fish and mussels. Along the Gulf Coast, crustaceans, shellfish, various sea life, acorns, hickory nuts and birds supplemented their diet. The Archaic Period, 8000 BC to 500 BC, brought about an enterprising age when humans probed their environment for useful foods and tools. One of the major tools to be invented during this period was the atlatl, or spear thrower. The atlatl was usually a short wooden shaft or antler with a notched protrusion on one end and a weight or banner stone in the middle. The notched protrusion fit into a nook on the shaft of the spear. The spear or dart was then placed on the atlatl. When thrown, it went farther and with greater velocity than hand-thrown spears. Stone tool production, stone atlatl weights, boat stones, plummets, ungrooved axes (celts), gorgets, grinding pallets, projectile points, smoking pipes, steatite bowls, mortars and pestles became more refined and increased during the Archaic Period. Late during the Archaic Period, fired clay objects appeared. Clay was used to manufacture cooking balls, ceramic figurines and clay pottery containers. The appearance of mound-building domesticated plants and horticultural practices ushered in a new age. The coastal people developed a unique coastal and river culture, with political and social organizations. They lived in larger villages or cultural centers.

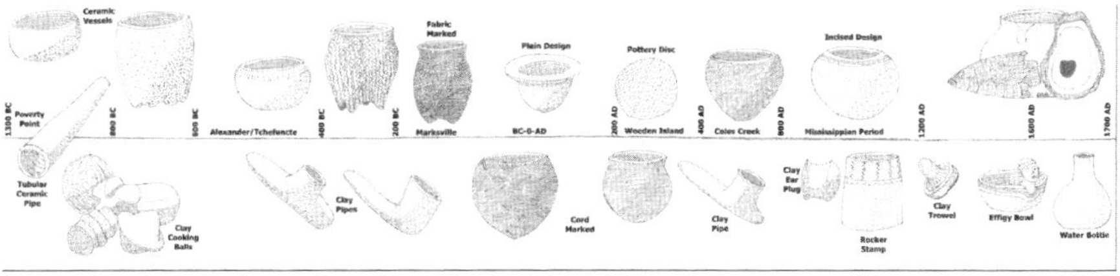

Timeline of Gulf Coast Native American pottery.

Many characteristics from the Archaic Period were carried into the early Woodland Period. Along the Gulf Coast, the Woodland Period lasted from about 500 BC to about AD 1000. The late Archaic and early Woodland Period along the southern coastal plains is sometimes called the Gulf Formational Stage. The presence of ceramics, burial mounds and increased horticulture indicates a new period. Ceramic vessels, which weigh more and are not readily transported from place to place, as well as increased cultivation of crops indicate more permanent settlements. Pottery became not only a useful everyday item but a beautiful art form as well. Stone effigies of animals and humans began to appear.

Compared to the earlier people, the Woodland people introduced a complex religious culture with burial mounds and various changes in mortuary practices. The mounds at first were simple earthen works raised over the dead, but with time they became more complex. Various archaeological excavations indicate the dress, ornamentation and grave goods of certain individuals, which suggest the creation of a social order. Construction of burial mounds took extensive human labor. Individuals of high status were sometimes buried with other humans who had been killed to accompany the deceased into the afterlife.

Villages and ceremonial areas apparently were well planned and laid out. Larger village complexes appear to have been supported by small villages. These small villages gathered essential goods and services from their region. These items, which were needed to sustain everyday life, were distributed between the two communities. In return, they received support and trade items from the large villages. Extensive long-distance trade routes were developed from the Gulf to the Great Lakes to procure exotic materials. Archaeological excavations have indicated that the cultivation of crops like maize, squash and sunflower took place during this period. Other native plants—the ones that had high yield and were hardy—were also probably grown. The Gulf Coast natives continued hunting animals and gathering sea life and plant life from the wild. The bow and arrow were introduced sometime during the early millennium AD.

The Woodland Period, with all its innovation, slowly gave way to the Mississippian Period. The Mississippian Period saw an approach to unification under one social culture or ceremonial complex. This society, which supported an expansive trade system, was based on a religious ceremonial culture centered on floodplain horticulture. The Gulf Coast had early water trade routes that extended up the Tombigbee River, to the Mobile area and farther east to the Fort Walton complex. This horticultural and trade economy helped support an art community and notable public works.

One characteristic of this period is the appearance of distinctly Mississippian pottery. The pottery not only had unusual designs, but along the Gulf Coast and its trade areas, it also contained crushed shell. While the clay was still in its soft form, crushed shells were added. Then the clay was shaped into a vessel and fired. Archaeologists call this process shell tempering. Vessels were made in all sizes and shapes for everyday use. Beautifully designed vessels, some shaped like humans, animals or birds, were designed for ceremonial purposes.

Large earthen works called mounds were built during this period. Some of these mounds were flat, with temples, upper-class houses or council buildings built on top. The common people and the slaves lived in houses at ground level near a central plaza. Wattle and daub houses were built with saplings or canes woven around the wall poles. The gaps were packed with a mud and moss mixture, to which a clay plaster was applied. The roof was constructed in the same manner but was covered with grass thatch. The buildings were rectangular or square in shape, with a centrally located fireplace. The fireplace provided heat in the winter and acted as an insect repellant in the summer.

The Mississippian people learned to use not only clay but bone, stone, shell, wood, canes and plant material as well. These materials were utilized as tools, cloth, rope and dyes and were manufactured into artwork and other utilitarian objects. Life here on the Gulf Coast was a system of hunting, gathering and growing crops. The hunters would hunt mammals and birds in the forest and harvest sea life from the Gulf waters and rivers using traps, nets, hooks, bows and arrows and even bare hands. They also gathered nuts, fruits and herbs; other crops were also grown.

The Gulf Coast has a large number of Mississippian sites. The earlier sites have a large number of artifacts that appear to be out of the original Mississippian Period homeland. This may indicate the early arrival of the

Mississippian Period pottery.

Mississippian culture to the Gulf Coast. The Mississippian Period lasted from AD 1000 to AD 1700 but had begun to disintegrate prior to the arrival of the Europeans in the 1500s.

The Historical Period began when the French arrived on the Gulf Coast in 1699 under the leadership of Pierre Le Moyne d'Iberville. From the translation of *Iberville's Gulf Journals* by Richebourg Gaillard McWilliams, we learn about the Native Americans. On February 13, 1699, Iberville and fourteen men went to the Biloxi peninsula to investigate smoke on an island that is present-day Deer Island. They tracked some Native Americans eastward on the Biloxi peninsula. This group of Native Americans consisted of members of the Biloxi tribe. Upon reaching the mouth of Biloxi Bay, they crossed over to Deer Island, where they joined up with several members of their tribe. They then fled in their dugout canoes up the Biloxi Bay, with the French in pursuit with a long boat and birch bark canoe.

The Biloxi tribe members fled into the woods upon reaching the north shore, leaving their dugout canoes, baggage and a sick old man. Iberville had the old man and a woman his men had captured, made comfortable and presented with gifts. On February 15, additional Native Americans came and sang the calumet of peace with Iberville. This began a long friendship between the Biloxis and the French. This friendship would later include the Pascagoulas, Capinans and Acolapissas.

The name "Biloxi" is most likely a corruption of the name the Biloxis called themselves, *Ta' neks Aya'di.* Iberville refers to the Biloxis as the Annocchy, whom the Bayogoulas called Bilocchy. They spoke a Sioux dialect quite similar to the Ofo and Tutelo with loose ties to the Dakota, Hidalsa, Mandan, Crow and Winnebago—all members of the Sioux family.

It has been suggested that the Biloxis migrated from the Northeast. There are indications that they arrived along the Mississippi Coast not long before the French. Their migration may have been due to European contact and disease. We do know that Europeans brought diseases to which Native Americans had no immunity. These diseases caused large-scale death in the native population. Iberville indicates that diseases had ravaged the Biloxis before 1699.

Iberville gives us the first description of the Biloxis and the Capinans on February 16. Iberville went to the location where the Biloxis and Capinans had spent the night. He described the Biloxis and the Capinans as having tawny skin and heavy, black, well-groomed hair. The group he talked to was made up of ten men with their weapons. All were entirely naked except for braguets. According to the French, a braguet was a loin flap for a man and an apron for a woman. One of the Biloxi warriors would guide French explorer Pierre Lemoyne Iberville on his first ascension of the Mississippi River.

At the time of the French arrival, the Biloxi, Pascagoula and Capinan villages were located on the Pascagoula River. In some of their documents, the French refer to the Capinans as the Moctobi or Moctoby tribe. Little is known of the Capinans, but due to their intimate relationship with the Biloxis, they may have been of Siouan linguistic stock. The Pascagoulas were of Muskhogean tribes and spoke Muskhogean dialect. The word "Pascagoula" is derived from the Choctaw *paska*, or bread, and *Okla*, or people, thus meaning "bread people." The Acolapissas were at the mouth of the Pearl River. The word "Acolapissa" is derived from Choctaw words that mean "those who listen and see." The Acolapissas were also part of the Muskhogean tribes. All of the Gulf Coast Indians probably spoke the Mobilian trade tongue.

Winter clothing consisted of braguets, belts, leggings, moccasins and skin robes. The women would wear animal skins or dresses made of woven fiber from the bark of the mulberry tree, which covered them from their waists down to their knees. It was a very hot day in August when the French came to visit. As such, the French give us additional information on summer attire of the Pascagoulas. In the *Fleur de Lys and Calumet* by Richebourg Gaillard McWilliams, a translation of the memories of Andre Penicaut, we find many descriptions of the Native American culture:

> *All the savages, the men and the boys, went as naked as one's hand; but the women and the girls wore a single hank of moss which passed between their legs and covered their nakedness, the rest of their bodies being quite*

> *nude. This moss is a very fine plant half an ill (40 inches) long, which the French in the region derisively name Spanish beard and the Spaniards, to return the favor, call French wig.*

The Biloxi, Acolapissa and Capinan clothing practices were most likely the same. The Biloxi ornaments included feather headdresses, and they were said to wear necklaces of bone and the bills of a long-legged red bird today known as the flamingo. Some tattooing was practiced, and nose rings and earrings were worn.

The Biloxis, Pascagoulas, Acolapissas and Capinans made pottery and wooden bowls, horn and bone tools and baskets. They produced their dugout boats by lighting a fire at the base of a cypress tree until the fire burned through the base, bringing down the tree. After all limbs were removed, a fire was then burned on top of the fallen tree to the desired length of the boat. Next, a fire was burned on top of the trunk until the desired depth was reached. The fire was doused with mud, and the trunk was scraped with large shells until the vessel was to the depth desired. One French drawing shows not only shells being used as scrapers but also the hipbone from a deer. Their weapons were bows and arrows, stored in otter, fox or other skins and carried on straps slung across their backs. They carried wooden clubs, which the French called head breakers, in their right hands. The French indicated that to strike a person's head with a head breaker was like hitting a watermelon with a stick because it had the same effect.

From the French accounts of the calumet or peace pipe ceremony, we know that their musical instruments included a chichitois, or gourd filled

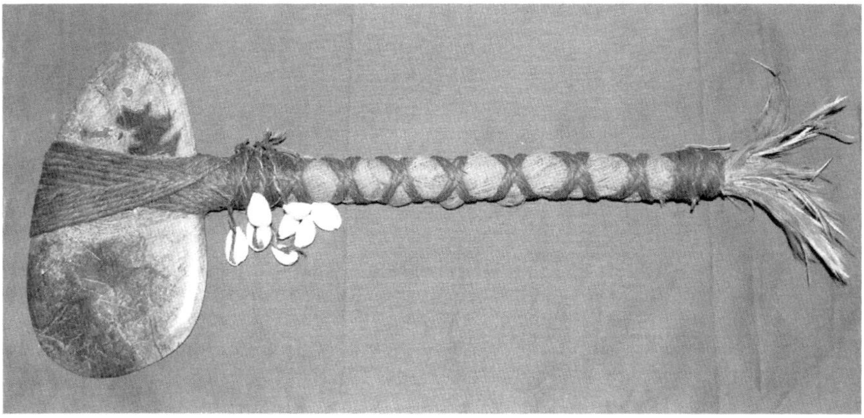

Reproduction of a Native American stone axe.

with pebbles and as large as two fists. A drum was fashioned from pots containing water and covered with a piece of deerskin stretched tight across the mouth.

The French also give some insight into the diet of the Mississippi Coast tribes. The women cooked in large bowls that held about forty pints and fed two or three families. Once again, from Penicaut:

> *They gave us something to drink and to eat such as buffalo, bear, and deer, and every kind of fruit in abundance, such as peaches, plums, watermelons, pumpkins and all of an exquisite flavor. The pumpkins, they are cooked without water, and the juice that comes from them is like syrup, it is so sweet. They served us also some of their sagamite, which is a kind of pap made from maize and green beans. Their bread comes from maize and from a grain that grows on canes.*

These tribes were dependent on small and large game, fish, shrimp, oysters and other shellfish. Their diet was probably supplemented with nuts, berries, herbs and other wild foods. The meat was smoked or dried on a wooden frame with a fire underneath until the meat was completely dried. They cultivated crops of corn and beans and possibly squash and sunflowers. Andre Penicaut described the coastal Native American process of planting:

> *The ground is prepared with flat, hooked sticks with which they pick the ground, as they know nothing about plowing the way we do in France. They scratch the ground with these hooked sticks and uproot the canes and weeds, which they leave on the ground out in the sun for two weeks or a month; and afterwards they set fire to them and when they are reduced to ashes, they take a stick as big as one's arm and sharpened at one end and make a hole in the ground every three feet, into which they put seven or eight grains of corn per hole and cover it over with dirt. When the corn is a foot tall they pull the weeds and continue to weed until harvest.*

The Biloxis, Pascagoulas, Acolapissas and Capinans proved to be extremely traveled groups during the French and Spanish colonial periods. The 1699–1702 locations of the Biloxis, Pascagoulas and Capinans were on the Pascagoula River near one another. In 1702, there were 250 Acolapissas living on the Pearl River. That same year, the Acolapissas moved to the north side of Lake Pontchartrain and migrated into various bayous in the area.

During the early 1700s, a group of Biloxis settled on a small bayou, St. Jean Bayou, between New Orleans and Lake Pontchartrain. Saint Denys, the commander of the first French fort on the Mississippi, could have caused this move because he made a request that the Biloxis move near the fort.

It appears that in 1708, a group of Biloxis moved to Fort de la Boulaye on the Mississippi to assist and bolster the French garrison. The year 1722 found some of the Biloxis settled on the Pearl River on the abandoned Acolapissa site. The Acolapissas were on the east side of the Mississippi River above New Orleans. French records of 1725 found the Choctaws making threats to cut the throats of the French and enslave the small nations. This induced the Mississippi Coast tribes to move nearer to Lake Pontchartrain for a short period of time. The census of 1725 showed the Biloxis and the Pascagoulas on the Pearl River to number about forty men. The Pascagoulas and the Capinans on the Pascagoula River numbered about two hundred men, but the Biloxis are not listed there. Between 1725 and 1730, some Biloxis appear to have wandered back to the Pascagoula River. Dumonte de Montigny, a French historian, places them near the Pascagoulas during this period. This may indicate that the Biloxis had divided into more than one group.

French records of 1738 show the Capinans had abandoned their village on the Pascagoula River and moved to Deer Island off the coast of Biloxi. This is the last record of the Capinans. They appear to have vanished from history or possibly were absorbed into other tribes. Some of the Biloxis were residing at Shell Island at the entrance of Lake Pontchartrain. This was done to remove them from surprise attacks of the Choctaws.

In 1758, Louisiana Governor Louis Billouart De Kerlerec indicates that the Acolapissas have been destroyed by the expansion of settlements and the liquor trade. Yet according to Claiborne's history of Mississippi, the Acolapissas united with the Houma and vanished from history. In December 1758, De Kerlerec indicated that some small nations such as the Chatots, Pascagoulas and Biloxis clearly obeyed their orders and did all sorts of services for the colonies. The Pascagoulas and Biloxis were still located on the Pascagoula River as late as 1763.

On January 30, 1764, Jean D'Abbadie, director general of Louisiana, reported that a delegation of Biloxis requested permission to settle on the right bank of the Mississippi River. The Biloxis pledged their loyalty to the French. They indicated that they wished to die among the French. This location was just off St. Jean Bayou in New Orleans. The Pascagoulas had also moved in Louisiana. In June 1764, D'Abbadie reported that the Biloxis and other small tribes had assembled near New Orleans to play ball and that the activity attracted many spectators.

During the 1766 census of Spanish Louisiana, which was conducted for Spanish Governor Antonio de Ulloa y de la Torre Guiral, both the Biloxis and Pascagoulas were listed. Spanish Governor Alexander O'Reilly listed the Biloxis and Pascagoulas among the tribes receiving their annual gifts. By 1771, some Biloxis lived on the Amite River, but due to Choctaw threats, they moved near the Pascagoulas on the Mississippi. That same year, John Thomas, British deputy superintendent of Indian affairs, indicated that the Biloxi numbered one hundred individuals. He indicated that they were living on the Mississippi River above Post Rapide near the Red River. In 1780, due to threats from British-allied tribes, some Biloxis moved to Avoyelles, while other Biloxis fled south to the Louisiana marshes. By 1796, some Biloxis and Pascagoulas had moved to Bayou Boeuf.

Dugout canoe unearthed by Hurricane Ivan in 2004 in Ocean Springs.

In 1784, British surveyors journeying up the Mississippi indicated that above Point Coupee, in present-day Louisiana, there was a village of the Pascagoula and, nearby, a village of the Biloxi. They indicated that the Pascagoulas numbered about twenty warriors, while the Biloxis numbered about thirty warriors. In 1787, some Pascagoulas relocated to the Red River. In 1795, the Pascagoulas moved to Bayou Boeuf, and below them were the Biloxis who had moved about a year before. In the early 1800s, the Biloxis and Pascagoulas sold this land on Bayou Boeuf to settlers.

In the territorial papers of Orleans territory, there is an account of the Indian tribes of Louisiana. According to this account, fifty Biloxis and Choctaws are listed as wanderers on Bayou Crocodile, which empties into Bayou Teche in south Louisiana. Sixty Biloxis are listed as living at Avoyelles and on Lake Avoyelles. One hundred Biloxis were living on Bayou Rapide. The Biloxis and Pascagoulas continued to live in this area, and at a later date, some would either die out or merge with the Tunica and Choctaw tribes.

In 1803, Spanish officials indicated bands of Biloxis in eastern Texas. In Louisiana, two Biloxi villages were reported near Avoyelles with 60 individuals. It was reported that in 1817, a large group of Biloxis migrated to Texas, to an area now called Biloxi Bayou, a branch of the Neches River in Angelina County. The same year, 100 Biloxis and Pascagoulas were listed together on the lower Red River. In 1822, a branch of the Pascagoulas was reported on Biloxi Bayou near the Biloxis, who numbered about 50 individuals. Two Pascagoula villages were listed on the Red River in Louisiana. Still, in 1829, records indicate that 65 Biloxis and 111 Pascagoulas were living on the Red River in eastern Texas.

As we have seen, the Mississippi Coast tribes were well traveled. Yet their journeys had just begun. Sometime between 1829 and 1849, most of the Biloxis living in Texas would eventually migrate (Indian Removal Act of 1830) to the Indian Territory in Oklahoma. The Pascagoulas most likely consorted with the Biloxis and moved to Oklahoma with them. Indian Territorial records list the Pascagoulas among the tribes living in Oklahoma. In 1849, a Biloxi village was listed near Fort Washita, Oklahoma, in the Indian Territory. This village was located in the area of the Chickasaws. More of the Biloxis were reported as having settled on the Kiamichi River and a few near Atoka.

The Biloxis and Pascagoulas also continued to maintain a presence in central Louisiana. They were described as sincere, inoffensive and friendly people who spoke a language all their own, as well as Mobilian, the trade tongue. In 1850, a group of Rapides Parish residents petitioned Congress for the removal of the Biloxis, but their petition was not successful. The

Biloxis were described as wearing buckskin breeches and calico hunting shirts, which they buttoned from their belts to their chins. Some wore brass wrist-, nose- and earrings. There was also a report that the Texas group would rendezvous with the Louisiana tribes and celebrate.

By 1886, Dr. A.S. Gatschet discovered a few Biloxis on Indian Creek west of Le Comple, Louisiana. Dr. Gatschet was with the Bureau of American Ethnology, and he began collecting and recording enough of the Biloxis' language to prove a Sioux connection. Between 1892 and 1893, Dr. J.O. Dorseu, a Siouan specialist with the Bureau of Indian Affairs, revisited this group and collected and recorded additional information on the Biloxi language and culture. His additional research proves Dr. Gatschet's original findings that the Biloxi spoke the Sioux language and were most likely Sioux. The 1900 U.S. Census of Indian Creek in Louisiana indicates just two individuals who claimed to be Biloxis. Eight individuals who claimed to be Choctaws also indicated that their mother was a descendant of the Biloxis. In 1908, two brothers living near Livingston, Texas, claimed their father was Biloxi and their mother was Pascagoula. By 1910, no one at Indian Creek in Louisiana claimed to be Biloxi or a descendant of the Biloxis. Today, east of Indian Creek area, the Tunica-Biloxi tribe resides in Marksville, Louisiana. Some of its members claim to be Biloxi or descendants of Biloxis.

With the arrival of the French in 1699, the four Mississippi Coast tribes embarked on an enduring alliance. These tribes shared all they owned and all of their knowledge with the French. If not for the Mississippi Coast tribes, the French would have died of starvation on the beaches of the Gulf Coast. In time, the Mississippi Coast tribes slowly began to merge and marry into other Native American tribes. These merges and marriages would eventually cause them to fade into history.

The French created alliances with the local Native Americans. They saw the earthen works and shell piles of the long-forgotten Paleoindian, Archaic, Woodland and Mississippian Periods but only wondered who had built them. Historical documents, archaeological excavations and artifacts recovered from these Native American periods are all that remain from these past civilizations that can help us understand the life and times of Mississippi Gulf Coast natives.

Chapter 2

Frenchmen, British and Spaniards

Colonial Biloxi's maritime history has connections to France, England and Spain. Trade along the ocean routes and the abundance of sea life were important aspects of the daily livelihoods of the colonists. Commerce was conducted throughout the Mississippi Sound and ranged from New Orleans to Mobile. In addition, foreign commerce steadily supplied European goods and exported North American goods.

Period French

In 1697, the Comte de Pontchartrain, French minister of marine, gave Pierre Le Moyne, Sieur d'Iberville, orders to locate the mouth of the Mississippi River. Iberville was a French Canadian and was selected due to his prior successes against the English along the East Coast and in his native Canada. In addition to locating the mouth of the river, he was tasked with building a post, defending the mouth of the river and blocking access to the river from the English and Spanish. His expedition left Rochefort, France, on September 5, 1698. Iberville's French flotilla anchored in the natural harbor of Ship Island on February 10, 1699.

On February 13, after seeing a campfire on an island, Iberville and fourteen men landed on the mainland at present-day Biloxi. Once on shore, they tracked the Indians eastward. After the first night, they continued eastward to the Bay of Biloxi. After reaching the bay, the Indians took their canoes and crossed over

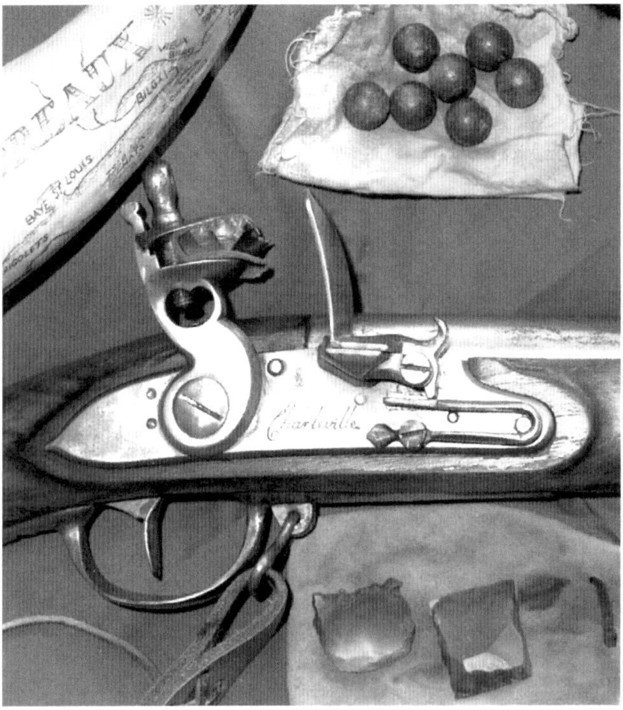

Reproductions of French musket showing flintlock with gunflint, sud and pan. Other items visible are round shot, gun flints and section of powder horn.

to Deer Island. On Deer Island, they joined other Indians and fled in several canoes to the north shore of the Biloxi Bay. Iberville pursued and overtook them at the north shore in either present-day Ocean Springs or St. Martin. The Indians fled into the woods, leaving canoes and baggage. After several days, the French befriended these Indians, known as the Biloxis. The arrival of the French ushered in a new period along the Mississippi Gulf Coast.

When Iberville located the mouth of the Mississippi River, he found it blocked by timbers and other materials. He decided to build a fort nearby. He knew that in a short time he would return to France, leaving behind a garrison to protect France's interest in this new frontier. He needed to construct a fortification to house the garrison. In April 1699, Iberville had his men construct Fort Maurepas, which was referred to as Fort Biloxi by the French. This fortification was located in present-day Ocean Springs. By 1702, the French colonial capital had moved from Biloxi (present-day Ocean Springs) to Mobile. The Louisiana French capital remained in Alabama until late 1719.

On October 26, 1719, the colony's governing body (known as the superior council) decided to relocate its capital back to Biloxi. The French reestablishment at Vieux (Old) Biloxi in 1719 occurred just prior to the arrival

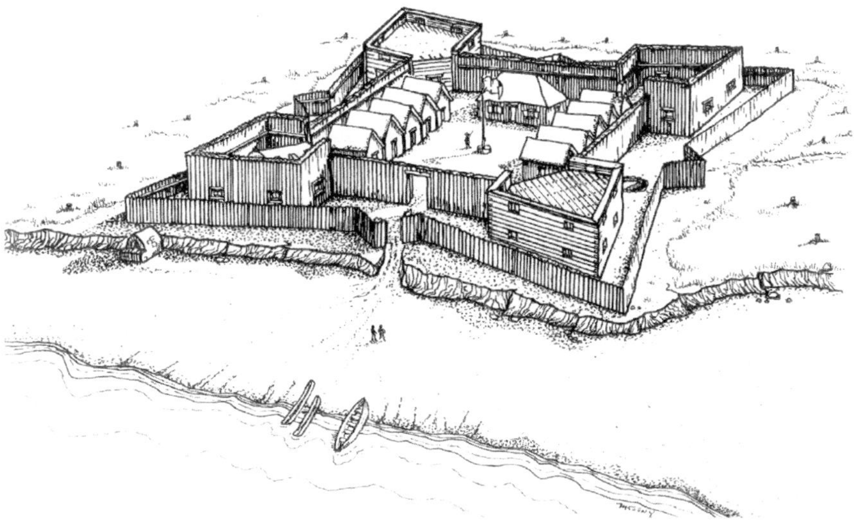

Fort Maurepas, built in April 1699 in present-day Ocean Springs. *Courtesy of the Maritime & Seafood Industry Museum.*

of a large number of colonists to the Gulf Coast and their recruitment by John Law's Compagnie des Indies.

John Law had been appointed treasurer of France by French King Louis XV. This occurred after Law proposed a system to make money for the king by selling shares in Law's Compagnie d'Occident. Law's Company of the West was created in August 1717. The French government gave the Company of the West the foreign trade rights to the Louisianan and Canadian colonies. By May 1719, the Company of the West had obtained the exclusive rights to trade with China and the East Indies. Law's company eventually controlled all trade outside of France and was renamed the Compagnie des Indies.

The colonists recruited by the Compagnie des Indies included many Germans and Swiss. The colonists worked for several concessionaires and land grantees in order to work off their land grants, but destiny brought them to New Biloxi. Some of the first colonists recruited by the Company of the West landed on Dauphine Island in late 1719. After their arrival, they were redirected to Ship Island and transported to Nouvelle (New) Biloxi (present-day Biloxi). A life and death struggle soon ensued along the shores of Biloxi.

In May 1720, 130 African slaves were transported to Louisiana. About 500 more African slaves arrived in 1721. A coastal Biloxi map dated April 23, 1722, shows that the houses of Negroes were located near the brickyard on

Biloxi's Back Bay. Jean Baptiste Le Moyne de Bienville, governor of French Louisiana and d'Iberville's brother, complained that most of the adults were old and sick. Shortly after their arrival, 28 of the sick slaves died.

The memoirs of Charles Le Gac, director of the Compagnie des Indies in Louisiana, indicate that in early 1721, about 2,129 individuals had debarked at New Biloxi. The concessions that lined New Biloxi beaches were those of M. Le Blanc, Minister of War Sieur John Law, Sieur de Meuse, Sieur de Chaumont, Sieur de Paris du Vernay, Sieur de Coly, Sieur Du Manoir, Sieur Villemont, M. Clerac and Sieur d'Artaguette. Each concession numbered about 250 individuals, which included foremen, technicians, artisans and laborers. Most of these concessions were to be located along the Mississippi River and its tributaries.

Each concession had to build its own boats in order to transport its workers to the land grants. They cleared land and built cabins and warehouses along Biloxi's shoreline. Some planted crops and began the construction of boats. By late 1720, the capital had moved from Old Biloxi to New Biloxi. New Biloxi was located in a ravine on the west side of Porter Avenue. Plans were drawn to build Fort Louis on that site. The fort's earthen works were started, but the fort was never completed.

On February 28, 1720, a squadron arrived at Dauphine Island from France. It was commanded by de Saujon and included the ships *L'Achille* (sixty guns), *Le Contert* (sixty guns) and *Le Mercure* (fifty-six guns). There were 208 passengers, including some young women taken from the public hospitals of Paris. They were each given a small trunk with a single wardrobe and sent to Louisiana to marry. Historians refer to them as the Casquette Girls. These young women were sent to Biloxi and rapidly placed in marriages.

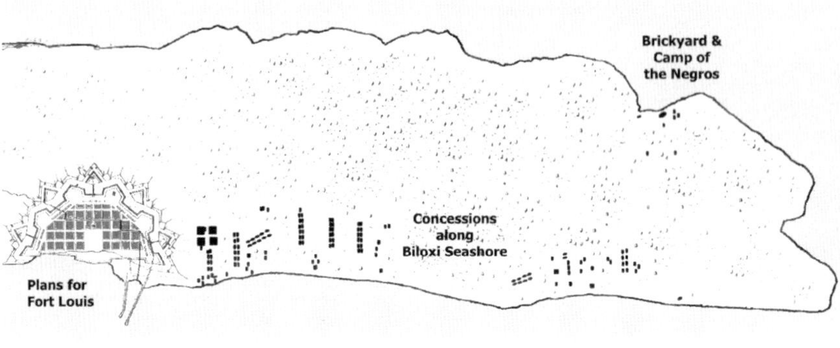

Map depicting plans of Fort Louis with French concession or land grants east of the fort and the camp of the Negros on Biloxi's Back Bay.

The period from 1720 to 1721 was extremely hard on the colonists. By 1721, 2,500 colonists and troops had been listed at New Biloxi. These colonists were living off company rations. This large number, the colonists' lack of knowledge of soil conditions and climate and the constant shortage of provisions created many hardships within the colony. It was reported that many of the colonists who arrived at Dauphine Island and Biloxi died from starvation or plague. The virtual famine forced many to eat plants and herbs that often caused death instead of nourishment. Many waded up to their thighs in the Mississippi Sound to coon oysters by hand. Records indicate that many got sick and died after eating oysters. It was noted that these deaths were due to the long time they were in the water. Today we refer to this as hypothermia. The famine was so great that Governor Bienville was obliged to send soldiers, workmen and officers to live with the local Biloxi, Pascagoula and other Native American tribes.

At the height of the famine, the priests were so busy that they could not keep up with death records. Between August and December 1721, 76 individuals died in New Biloxi. The Du Manoir concession, which consisted of 250 individuals, lost 73 people between 1720 and early 1721. French colonist M. Le Page Du Pratz of Natchez wrote in his *History of Louisiana* that "the famine being there [New Biloxi] so very great, that more than five hundred people died of hunger."

French cannons recovered from Biloxi Bay in 1892 by Eugene Tiblier Jr. The cannons are on display near the Biloxi small-craft harbor.

The arrival of provisions in 1721 brought the famine to an end. New Biloxi rolls placed the 1721 population at about 2,500, and some historians estimate the number of dead from the famine to have reached the thousands. This is not likely, due to the fact that the colony survived. Du Pratz places the number at about 500, which seems more plausible. The loss of 500 individuals would have been a huge blow to the colony. Between 1717 and 1721, about 7,020 individuals had been transported to French Louisiana. Of this number, about 2,000 had died, deserted or returned to France. This left the European population in the French Louisiana territory at 5,420 individuals in late 1721.

By 1722, Governor Bienville began his efforts to move the colonial capital to New Orleans. In October, Bienville reported that ships drawing thirteen feet of water had entered the Mississippi River. On February 1, 1723, the capital of Louisiana was officially moved from Biloxi to New Orleans. A garrison of six men, commanded by a sergeant, was detailed for the fort at Biloxi. Biloxi continued as a garrison post until 1730, at which time the garrison moved to New Orleans.

French settlers were located near the mouth of the Pascagoula River, the Bay of Biloxi, the Bay of St. Louis and the Pearl River. The year 1755 witnessed the beginning of the French and Indian War between France and England for

French colonial reenactment of French arrival in the village of the Natchez Indians, Natchez, Mississippi.

control of North America. France fared poorly in this war. At the signing of the Treaty of Paris in 1763, France ceded all of Louisiana east of the Mississippi (with the exception of the Isle of Orleans) to England. The territory west of the Mississippi and the Isle of Orleans went to Spain. These concessions ended the French attempt to settle Louisiana. The French plan to build a chain of strong fortresses from Canada to Louisiana halted. The plan failed in part because of a lack of settlers and of the inability of French missionaries to convert large numbers of Indians. The French planned to rely heavily on Indian allies who could supply a huge army during time of war. French dreams of obtaining silver, gold, pearls and other resources had also failed; little had been realized. Biloxi and the Mississippi Gulf Coast were on their own.

British Period

The Paris conference of 1763 marked the end of the war and the beginning of British rule along the Mississippi Gulf Coast. In addition to the Mississippi Gulf Coast and the ports of Mobile and Pensacola, everything on the east side of the Mississippi River (with the exception of the city of New Orleans and surrounding peninsula) fell under British rule. The treaty also opened the Mississippi River for use by Spain and Britain.

The first British troops ordered to Pensacola were the Twenty-second and Thirty-fourth Regiments under the command of Major Robert Farmar and the third battalion of the Sixtieth Regiment under Lieutenant Colonel Augustin Prevost. These first troops were veterans of the Havana campaign of 1762. It was during the course of this campaign that they were razed by disease, battle and the climate.

The British initially placed troops at Pensacola and Mobile. In time, posts were created at Natchez and Baton Rouge and along the Iberville and Tombigbee Rivers. In short order, they began exploring and surveying the northern Gulf of Mexico and the Mississippi River. Lieutenant Thomas Hutchins was sent by General Thomas Gage to British West Florida to survey, make maps and strengthen the fortifications at Pensacola. During his surveys and mapmaking, he described his findings in his logbook. Along the Mississippi Gulf Coast, he began his survey at the Pearl River and described it in great detail from the Pearl River to the Pascagoula River.

Lieutenant Hutchins reported that the Pearl River ascended to the lands of the Choctaw nations and was navigable to 150 miles from its mouth. There were some settlers on the river in 1769. These inhabitants had settled

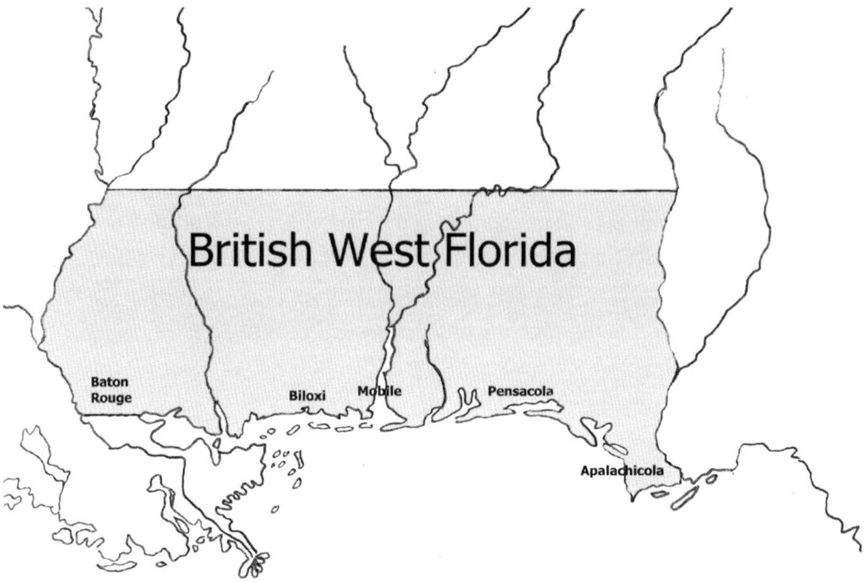

Map of British West Florida, 1763–80.

on the bank of the western branch of the Pearl River. The land on the opposite bank was used for their rice plantations. One of the settlers was M. LaRonde, who had a plantation about 15 miles above the mouth of the Pearl River. Above LaRonde were LeFaber, Vincent, De la Gautrais and Nieri. In addition to rice, these settlers raised tobacco, indigo, cotton, corn and various other vegetables.

As Iberville noted in his journal, Hutchins indicated the timbers of the coast area as excellent "for pipe and hogshead staves, masts, yards, and all kinds of planks for ship building." Another British surveyor, George Gauld, noted in 1768 in his rough field charts that the "Pearl River abounds with excellent timber for shipbuilding and all kinds of mast, yards, etc."

Upon leaving the Pearl River, Hutchins journeyed about eighteen miles to the Bay of St. Louis. He found the bay very beautiful but at the same time compact, and he thought that the land was good for pastures with light soil. He indicated seven feet of water at the entrance of the bay, with the water deepening as one moved farther along. He reported the land as not having been occupied since 1767. Choctaw Indians led an attack against the settlers that year and drove the settlers away by killing all of their cattle.

Hutchins then journeyed twenty-six miles to Biloxi Bay. He reported that Old Biloxi was situated north of Ship Island with a small bay just behind

Deer Island. He indicated that this was the first establishment of the French in Louisiana. He found inhabitants of the area whom he believed were descendants of the original French settlers. He found them making a living by raising cattle and other animal stock and making pitch and tar. The Indians seemed to be pesky, but the settlers were hanging on.

Hutchins continued his journey to the Pascagoula River, about thirteen miles away. Here he reported that the river has several mouths between the western and eastern shores. The western mouth has the only true channel, which is four feet deep. He indicated that twenty miles upriver, the main river divides and continues its course to the Gulf. The area between the two branches has nothing but marshes intersected by bayous.

Hutchins reported that once above the mouth of the river, there are three to six fathoms of water. His reports indicate that the river was navigable for over 150 miles and that the soil of West Florida improved as he traveled up the river. He also noted that the coast-area soil was not bad either. On the east and west sides of the river, there were good plantations, but as in Biloxi, the colonists were harassed by the Indians trying to kill their livestock.

On the west side of the Pascagoula River, the plantations of Baudreau dit Graveline and Pierre Paquets were located near the present-day city of Gautier. George Gauld, British surveyor, surveyed the Pascagoula journey up its east branch and into Bayou Casotte. He also visited the plantation of Hugo Krebs.

Hutchins took time to describe the islands of the Mississippi Sound, beginning with Cat Island. He reported that Cat Island was about eight miles east of the Bay of St. Louis. It was situated seven miles from the mainland and was six miles long. The soil was meager, with only pine trees, some live oak trees and grass growing. The shore was blanketed by a large quantity of shells. Ship Island was about nine miles long and two miles at its widest width. There were large stands of pine tree and grass, but the west end was void of trees. The water from a well on the island was bearable. A one-mile-wide natural channel existed on the east end. He reported that the harbor afforded excellent protection from the elements during the summer but not during the winter. He also remarked that Ship Island was an excellent port, being easily accessible from Spanish Louisiana and the surrounding areas.

He reported that Horn Island was about seventeen miles long, with a large stand of trees in its middle section. The British also indicated that an abandoned old hut and an old house were located on Horn Island. The east end was void of trees but did have a large number of sand dunes.

A small key known as Dog Island existed between Horn and Ship Islands. He described the island as being surrounded by a shoal. Coast mariners quickly recognized these shoals as Dog Keys, and the island was later known as the Isle of Caprice. He noted that Round Island had a large number of trees and was about five miles north of the center of Horn Island. It was also reported that an old abandoned house was located on Round Island.

British colonial troops and defenses were never very strong in British West Florida, but they realized that West Florida could give them easy communication and trade with New Orleans by way of Bayou St. John from Lake Ponchartrain. About six miles up the bayou was a landing place where vessels loaded and unloaded within two miles of New Orleans. Creating commerce with Spanish Louisiana would be beneficial to the British. Pensacola had become a trading post for furs and skins. The British were receiving very large quantities of pelts. Mobile was also trading skins and furs, which amounted to about £12,000 to £15,000 sterling each year. These furs were being exported to London and various North American areas like South Carolina. Even though the fur trade was faring well, other trading ventures had failed. Hutchins indicated that Indian trade, indigo, cotton, rice, hemp, tobacco and lumber were the natural products of this country and should have been pursued and developed into trade. Fruit trees such as oranges, figs and peaches grew well and should have been researched.

During Hutchins's journey up the Mississippi, he noted that above Point Coupee in present-day Louisiana was a village of the Pascagoula Indians, and nearby was a village of the Biloxi Indians. He indicated that the Pascagoulas numbered about twenty warriors and the Biloxis numbered about thirty warriors. He noted during his explorations that crawfish were abundant in present-day Louisiana. The inhabitants had a small pond near their gardens filled with crawfish, which they harvested for their own pleasure.

During the American Revolution, Bernardo de Galvez and Spain secretly aided the American cause. However, in April 1777, Galvez directly attacked British merchants by seizing eleven vessels and their merchandise at New Orleans. On March 4, they reached Pensacola, and the HMS *Sylph* and HMS *Hound* were sent to New Orleans. There they aimed their guns on the town while the town's guns faced away from the river. Neither side wanted war, so Galvez agreed to release the vessels and British property in a timely manner. This satisfied the British merchants, and the British naval force withdrew and returned to Pensacola.

In January 1778, American rebels began to raid British settlements along the Mississippi. In February, raids occurred in the Natchez and Manchac areas. Due to these raids, the British Sixteenth Regiment arrived on the Iberville River to again take possession of Post Manchac. By January, the Waldeck regiment had arrived to strengthen the British post. In short order, Galvez set into action his plan to drive the British out of West Florida. His forces attacked and overran Post Manchac, taking its occupants prisoner. On September 21, 1779, both Baton Rouge and Natchez surrendered. The British troops at Natchez, due to the agreements of their surrender, were allowed safe journey to Pensacola. At the Battle of Baton Rouge, 350 British troops were captured.

Galvez then turned his attention to Fort Charlotte at Mobile. In 1780, his force attacked Mobile and forced the city to surrender within two weeks. Galvez set his sights on Pensacola but knew that it could only be attacked from the sea. Due to the British naval power at Pensacola, Galvez had to plan his next move carefully. In early 1781, the British Royal Navy was beginning to weaken in the Gulf due to needs in other theaters of the war. A Spanish flotilla was assembled and arrived off the coast of Pensacola on March 9, 1781. The siege lasted for two months, and on May 10, British forces at Pensacola surrendered. To ensure that British forces would not return, Galvez took the opportunity to send all of the British military and most of the British civilian population to New York. The fall of Pensacola gave Spain complete control of West Florida. Five months after the surrender of Pensacola, the British surrendered to the Americans at Yorktown.

Spanish Period

The Treaty of Paris was signed on September 3, 1783, and resulted in the British ceding Florida to Spain and the United States receiving its independence. Spain placed West Florida, including the Mississippi Gulf Coast, under the jurisdiction of Louisiana. All of these events had little or no effect on the inhabitants of the Mississippi Gulf Coast. Spain issued orders honoring the land claims of the existing families, and at the same time, Spanish officials were very liberal with new grants. In 1781, Spanish officials issued three land grants on the Biloxi Peninsula. James Mathurin, the widow Baptiste Christian and Nicholas Carceaux were recipients of these early grants. During the Spanish period, the mostly French families along the Mississippi Gulf Coast were not involved very much in land litigation. Nor would the Spanish show any interest militarily or politically until 1804.

The Spanish Crown promoted Galvez to field marshal of the royal armies due to his military success in West Florida. This promotion resulted in a new governor being appointed. Exteban Miro was appointed governor general of Louisiana and West Florida in 1782. He inherited military forces that can only be called minimal for the vast area of Louisiana and East and West Florida. The 1784 military count indicates that the Louisiana Infantry Regiment only consisted of two battalions, with each consisting of eight companies and an artillery corps. During emergencies, Mexico's and Cuba's militaries would supply troops and support. After taking over, Miro began the reorganization of the military and added more infantry, artillery, cavalry and militia to Louisiana and West Florida.

The Spanish government adopted new immigration policies in Louisiana and West Florida on July 14, 1787. These policies were created to relax Spain's strict rules on the economy, government and religion. This was done to attract other settlers into Spanish Louisiana and West Florida, particularly the Americans. Yet free enterprise was not the intent of Spanish economic policy or procedures; rather, the intent was to support the needs of Spain. Spanish government officials regulated the price and quality of goods. These officials burned inferior produce, offered subsidies and franchises and regulated commerce.

The following individuals are listed as living along the Mississippi Gulf Coast in 1791. Bay St. Louis (this included all the land surrounding the bay) consisted of Francisco Carriere, his wife and twelve children, as well as a free black named Mongula and a free mulatto named Peti Jean. On Cat Island there was Nicolas Christian, his wife and ten children; Jose Morin and his wife; two free blacks named Jose Lucer and Jose Mirer; and a free mulatto named Diequir Duret. On the Biloxi Coast (this included areas west and north of Biloxi Bay and parts of present-day Gulfport) were Luis Bourg, his wife and two children; Bautista Lanzete, his wife and five children; Luis Lancete, his wife and four children; Nicolas Calcaut, his wife and six children; and Santiage Ladner, his wife and five children. Free blacks and mulattos were Henrique Alexandro, Ladeosener, Bernardo, Francisco Mami, Luis Duran, Catini, Augusto Colin, Catalina and Honore Colin.

Along the Pascagoula River (which included land from Ocean Springs to Pascagoula and along the river system) was eighty-four-year-old widow Maria Bodro and her child; Jose Creker, his wife and three children; Francisco Creker (widow) and her two children; Augustin Creker, his wife and five children; Creker, a thirty-five-year-old widower with eight children; Gargare, a fifty-nine-year-old widower with three children; Santiago Quian and his wife; and Carlos Elies, his wife and five children. Free blacks and mulattos were Rosa Ysidro Duret, Toinete Edobaro Duret, Maria Josefa, Alexis Trullet, Manete Trullet, Noel Trullet, Maria Amon, Juan Bautista Amon, Juan Grena Landry, Romeno Landry, Ysabela Landry, Maximiliano

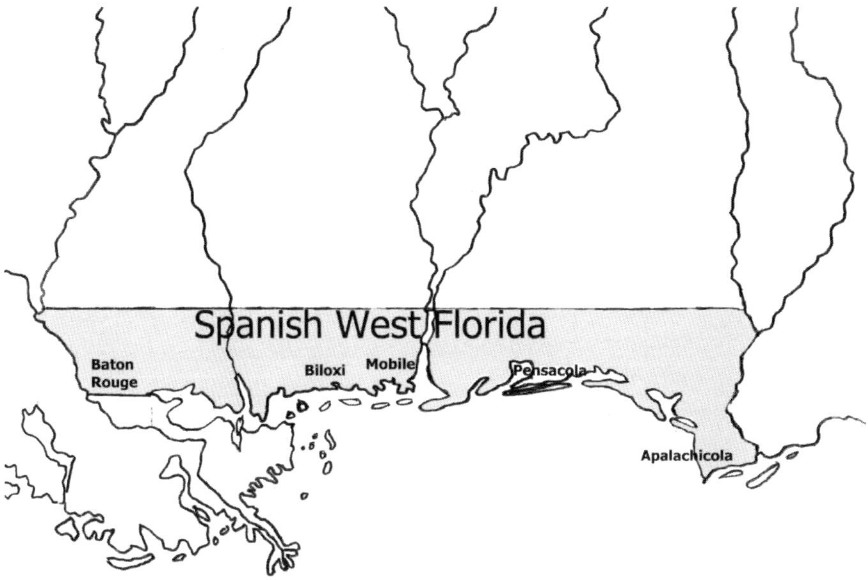

Map of Spanish West Florida, 1780–1810.

Landry, Juana and Francisco. There were twelve slaves on Carriere land in the Bay of St. Louis and forty-six listed in the Pascagoula River area.

Spanish Governor Miro was relieved of his command by Francisco Luis Baron Hector de Carondelet on December 30, 1791. Many small skirmishes occurred between Spanish forces and various groups from the United States. The United States and Britain entered into Jay's Treaty on November 19, 1794, and this placed pressure on Spain to open navigation of the Mississippi River. The Treaty of San Lorenzo, also called Pinckney's Treaty, between the United States and Spain was signed on October 27, 1795. This gave the United States access to the Mississippi River south of the thirty-first parallel, and Spanish control of Natchez District ended. Spain relinquished land between the Mississippi and Chattahoochee Rivers and lands above the thirty-first parallel to the United States.

Spain and France signed a secret treaty on October 1, 1800. The Treaty of San Idefonso de la Granja would cede Louisiana to France for some European concessions. Between September and December 1801, the United States entered into negotiations with France to acquire Louisiana. On September 15, after reviewing the extensive details of the negotiation, Napoleon agreed to sell Louisiana to the United States. King Charles IV decreed that Louisiana be retroactively ceded to France on October 15, 1802. The United States then purchased Louisiana from France for $15 million on May 2, 1803. By these agreements, the Louisiana Purchase is retroactively dated to April 30.

On October 31, 1803, President Thomas Jefferson named William Charles Cole Claiborne temporary governor of the Louisiana Territory with powers over the Spanish government. On November 20, 1803, Salcedo transferred Louisiana to Pierre Clement de Laussat, prefect and Napoleon's representative. Then on December 20, Laussat delivered Louisiana to Governor Claiborne.

Claiborne claimed that West Florida from Perdido River westward was part of the Louisiana Purchase. Spanish officials indicated that West Florida was not included in the original purchase. Due to these events, President Jefferson had Fort Stoddard built above Mobile on the Tombigbee River. Very few inhabitants along the Mississippi Gulf Coast were aware of the struggles or the confusion over ownership of West Florida.

Mississippi Gulf Coast inhabitants began to ignore Spanish orders and refused to perform their duties by serving in the militia. Americans living in other areas of West Florida became dissatisfied with Spanish rule. If the United States could not annex West Florida, they would fight for independence. The rebellion began with the occupation of Bayou Sarah in present-day Louisiana in early September. On September 22, forces of the rebellion entered Baton Rouge, which fell in short order. A temporary government was put in place by the insurgents. This was followed by an election of delegates that would attend a general convention to declare the will of the people. Delegates assembled in September, with John Rhea as chair and Dr. Andrew Steele as secretary. They declared their independence under the name of the Commonwealth of West Florida.

The new country, in a formal notice to President James Madison, recorded their actions and solicited annexation as an equal partner in this territory. They adopted a white star on a blue field as the republic's flag. President Madison would make no formal reply. He would later in a proclamation state that West Florida was part of Louisiana and that it had been purchased from France in 1803.

On October 27, 1810, President James Madison issued another proclamation ordering William Charles Cole Claiborne, governor of the Louisiana territory, to occupy the territory and invite the inhabitants to respect Claiborne, to obey the laws and to preserve order. Claiborne was to assure them that they would be adequately protected. The president issued this proclamation to prevent foreign intervention and to maintain order between the opposing factions. Claiborne took command of the area from Baton Rouge to a point just short of Mobile. For the time being, Mobile and

Pensacola were left to Spanish rule. This left the inhabitants of the Mississippi Gulf Coast with no administration, no laws and no officers. At the same time, they were still part of the independent Republic of West Florida.

The independent country known as the Republic of West Florida lasted only seventy-four days. Between December 1810 and January 1811, Governor Claiborne sent troops to the Bay of St. Louis, thus beginning the enactment of President Madison's proclamation. By the end of January 1811, the Mississippi Gulf Coast was under United States jurisdiction.

In 1811, Dr. William Flood, acting as ambassador for Governor Claiborne of Louisiana, investigated the Mississippi Coast for the United States. Dr. Flood left New Orleans aboard the USS *Alligator* and arrived on the Pearl River at the home of Simeon Favre. He commissioned Favre as justice of the peace for the Parish of Biloxi. He journeyed from the Pearl River to Bay St. Louis and Pass Christian, where the American flag was raised. Dr. Flood commissioned Philip Saucier as justice of the peace for Bay St. Louis. At Biloxi, Dr. Flood commissioned Jacque Ladner as justice of the peace for the Parish of Biloxi. He indicated in his report that the population of Biloxi was 420 people, chiefly French and Creoles. He described these early Biloxians as "a primitive people, of mixed origin, retaining the gaiety and politeness of the French, blended with the abstemiousness and indolence of the Indians." They depended on fish and game while planting rice and vegetables as a supplement.

Dr. Flood continued his journey to Pascagoula, where he commissioned Mr. F. Coming, Captain George Farragut and Benjamin Goodin as justices of the peace for the Parish of Pascagoula.

In January 1812, Governor Claiborne and George Poindexter, territorial delegate of the Mississippi Territory, agreed to extend the Mississippi Territory below the thirty-first parallel. The Mississippi Territory gained an area that extended to the Gulf of Mexico, with the Pearl River serving as the western border and the Perdido River as the eastern border. This agreement made the Mississippi Gulf Coast part of the Mississippi Territory under the governorship of David Holmes. Governor Holmes was appointed governor of the territory in 1809 and served until 1817.

Mobile continued to be held by the Spanish until January 12, 1813, when General Wilkinson was authorized by Congress to take control of the rest of Spanish West Florida. On March 8, United States naval forces under the command of Commodore Shaw requested surrender from Colonel Constant, Spanish commandant of Fort Charlotte Mobile, and the Spanish peacefully retired from Mobile. With little or no fanfare, the colonial period faded into the history books.

Chapter 3

Queen of the Watering Holes

In 1817, Mississippi became a state. Between 1817 and 1860, Biloxi began the slow development from a primitive area into a favorite summer resort for many southerners. In 1838, Biloxi was incorporated as a township.

During the early 1800s, boats were vital for Biloxi's existence. From the 1700s to the early 1800s, sailing vessels brought vital supplies to Biloxi. In the late 1700s, the use of steam engines to power ships had begun. By 1787, the steamboat era in America had arrived. In 1811, the steamboat *New Orleans* was operating on the lower Mississippi River. Steamboat service on the Mississippi River would bring plantation owners to New Orleans. In the Mississippi Sound, steamboat service was established from New Orleans to Mobile. This service would soon stop at various locations along the Mississippi Gulf Coast, bringing and loading goods destined for New Orleans or Mobile. Soon, people were traveling by steamboat, and visitors began to arrive. Many of the wealthy of New Orleans and southern plantation owners had summer homes in Biloxi. New Orleans's citizens traveled to Biloxi and the Mississippi Coast in hopes of escaping the yellow fever epidemics. Many of New Orleans's wealthy built large waterfront homes. When they traveled by steamboat to Biloxi, they would bring entire households, servants and all. These servants were usually household slaves. Some plantation owners from Mississippi, Alabama and Louisiana would also build summer homes.

By the late 1850s, Biloxi was considered "Queen of the Watering Holes," with several fine hotels or watering places, as they were called. They included the Magnolia Hotel, American Hotel, Biloxi House, Green Oaks Hotel and

The Riviera Hotel front beach was destroyed in Hurricane Camille. The Riviera was originally named the Shady Oak when it was built in the 1800s. It would later become the Montross Biloxi and finally the Riviera. *Courtesy of Alan Santa Cruz Collection.*

Shady Oaks Hotel, as well as many fine boardinghouses. The Shady Oaks in later years would become the Montross and later the Riviera. Tourism had become a new industry along the Mississippi Gulf Coast, and its future looked promising.

During this period, most of Biloxi's black Americans were slaves, but there were some free blacks in the Biloxi area. Many black slaves were household slaves for the wealthy home owners, while others worked in sawmills, brickyards or hotels. The free blacks owned their own farms, businesses and boats. One of the free blacks was a man named Bernard who was a blacksmith by trade. He was well respected, and Bayou Bernard in the Biloxi Bay was named for him.

Immigrants to Biloxi came from Italy, Ireland, Germany and France, and there were also people of English, Scottish or Scotch-Irish ancestry. Some of these immigrants arrived in the Port of New Orleans, while others arrived from neighboring states. They came and worked in the sawmills, turpentine industry, boat yards, stores, brickyards and hotels.

Boat building in Biloxi and North Biloxi had been practiced since colonial times. During the 1800s, many individuals built their own boats. Over time, master boat builders and shipyards began to appear along Biloxi Bay. Boat building skills, knowledge of repair and being an experienced boater were a

A Biloxi shipyard with schooners on the ways and at anchor on Biloxi's Back Bay. *Courtesy of Alan Santa Cruz Collection.*

necessity along the Mississippi Gulf Coast. Due to the Mississippi Sound's shallow waters, boats had to be specially designed for these waters. Schooners with centerboards that could be raised and lowered were utilized.

As water commerce and travel increased, problems of safety also increased. This brought about the need for local lighthouses. During a period from 1831 to 1889, lighthouses were constructed along the Mississippi Gulf Coast. The Cat Island lighthouse was constructed in 1831, the Ship Island Lighthouse in 1853, the East Pascagoula River Lighthouse in 1854, the Round Island Lighthouse in 1859, the Horn Island Lighthouse in 1874 and the Lake Borgne Lighthouse in 1889.

On March 3, 1847, Congress authorized the building of a lighthouse in Biloxi at a cost of $12,000. The metal sheets were cast by Murray and Hazelhurst Vulcan Works of Baltimore, Maryland. The lighthouse's cast-metal sheets were shipped in sections on schooners and assembled in Biloxi. A brick foundation was placed, and the cast-metal sheets were bolted together. Once the cast iron was assembled, brick layers were placed in the interior to create a brick liner. The work on the Biloxi lighthouse was completed in the spring of 1848. The cast-iron lighthouse stands sixty-one feet tall. Inside the lighthouse there are fifty-five steps on the circular stairway and a ladder to cover the last few feet. The first lighthouse keeper was Marcellus J. Howard.

A 1900 postcard of the Biloxi Lighthouse, with the lighthouse keeper's home seen behind the iron cast lighthouse. In 1941, the City of Biloxi purchased the lighthouse and lighthouse keeper's home. *Courtesy of Alan Santa Cruz Collection.*

Between 1828 and 1848, four of the lighthouse keepers were women: Mrs. Mary Reynolds, Maria and Miranda Younghans and Edna Holley. These women would serve for seventy-four years.

In 1850, the Mississippi legislature passed a law that established the Geological Survey of Mississippi. In August 1852, Benjamin L.C. Wailes conducted agricultural and geological surveys along the Mississippi Gulf Coast. When Wailes arrived in Biloxi, he took lodging at Mrs. Nixon's boardinghouse. The boardinghouse, called the Nixon Hotel, was on the corner of Nixon Street and Beach Boulevard. In 1848, one New Orleans newspaper describes the Nixon Hotel: "The beauty of the location, the extent of the improvements, the lavish expenditure for attentive servants, recherché viands and epicurean liquors, and, above all, the supervisory care of the intelligent and experienced matron conspire to make, Nixon's Boarding House the best watering hole of the place."

After visiting with Mrs. Nixon, Wailes next paid a visit to an individual he called an "old sailor and fisherman." This man, Charles, was noted for collecting fish specimens for naturalists, and he presented Wailes with two moon fish, a hog fish and a small turn, or turning fish, for his specimen collection.

Wailes noted that Deer Island sat in front of Mrs. Nixon's place and was about a half mile from the shoreline and twelve miles from Ship Island.

Wailes would also indicate that the weather for the past two days had been overcast with light showers. Some local fishermen reported that a gale had been blowing in the Gulf.

On Wednesday, August 25, 1852, Wailes reported that the previous evening, rain and heavy winds had begun to blow from the north, and by morning winds had increased to hurricane force. The hurricane continued all day Wednesday, and Wailes indicated that chinas and other similar trees were uprooted. The live oaks and magnolias lost limbs and appeared to be maimed. Numerous fences and some small houses were damaged or destroyed. In Mrs. Nixon's two-story building, the boarders were forced to flee the structure. The building was shored up to prevent it from being blown off its brick piers. Reports of the 1852 storm indicate that the eye went ashore somewhere between Pascagoula, Mississippi, and Mobile, Alabama, the evening of August 25. Damage from this storm extended from Pensacola, Florida, to Bay St. Louis, Mississippi.

While in Biloxi, Wailes visited Dr. Andreas Byrenheidt's office. Dr. Byrenheidt was a local physician. Here, Wailes met Judge C.P. Smith, Mr. Rivers, Mr. John Watt of New Orleans and Dr. Byrenheidt. Wailes indicated that Dr. Byrenheidt had been a surgeon in Napoleon's army. After leaving France, he resided in Point Coupee, Louisiana, before coming to Biloxi.

Wailes went riding on Back Bay, which he called "the extension of the Bay of Biloxi." He reported that he crossed over on a steam ferry. This steam ferry was operating from the Lameuse Street Warf on Back Bay. On the North Biloxi side, he visited the brickyard of Mr. George Kendall. Wailes described "the extensive brick yard of Mr. Kendall, where bricks are made on a very extensive scale from dry compressed earth, by steam power." Kendall brickyard was located in present-day D'Iberville. Wailes also recorded, "A number of small craft was in the Bay, and several along the shore undergoing repairs. Several steam mills, which are very numerous on the Bay, for sawing pine timber, were also in view."

On Saturday, August 28, Wailes awoke at dawn and journeyed to the market. Here, he purchased more fish and filled a four-and-a-half-gallon jar. He wrapped each fish individually and then filled the jar with alcohol. After cementing the jar lid, he placed it in a strong box, which Mrs. Nixon agreed to ship for him. His total expense at the Nixon Hotel was $19.85 for five days. After breakfast, he began his journey to his home in Natchez.

In the mid-1800s, tourism in Biloxi was good, with only a few hurricanes causing problems. But a different kind of storm was brewing, this one between men. The storms of disparity had been brewing between the

Biloxi City Hall and market, 1890s. *Courtesy of Alan Santa Cruz Collection.*

industrial North and the agricultural South. In 1861, Mississippi seceded from the United States, and the Civil War had begun. The seeds of civil war had been sown and in due time would affect the Mississippi Gulf Coast. The Southern states formed a confederacy, which led to the formation of military and home guard units. General David Twiggs from Pascagoula was placed in command of Louisiana, the coast of Mississippi and Alabama. During the Civil War, the Third Mississippi Infantry CSA was made up of Mississippi Gulf Coast natives. These were the Biloxi Rifles, Shieldsboro (Bay St. Louis) Rifles, Gainesville Volunteers, Dahlgren Guards of Pass Christian and Live Oak Rifles of Ocean Springs.

Fort Massachusetts on Ship Island was taken over by the Confederate forces. Afterward, Federal ships began to shadow the island and the Mississippi Gulf Coast. The Confederate forces held the island until September 16, 1861, when they were ordered to evacuate. Federal forces were now in command of Ship Island, and a blockade of commerce had begun. Many incidents and landings would occur along the Mississippi Gulf Coast, but there were only a few skirmishes between the two forces.

Due to the Federal blockade, many businesses suffered, and some would go under. Many families were in for hard times and sometimes went hungry. One fish that was plentiful and easy to catch was the popeyed mullet. Soon it became a mealtime staple and became known as "Biloxi Bacon." Even though the woods were full of game, ammunition was scarce; therefore, many people raised corn and chickens to live on. Salt during this period was instrumental in preserving meats and other staples but was one of the items that were very scarce. Many families would remove the salt from salt water by the process of evaporation. Today, we call this sea salt.

Both Confederate and Union documents indicate a battery location near the lighthouse. The Union forces, in fact, kept a good distance from this area due to what they thought to be a formidable battery. But all of that would change on December 31, 1861, when Lieutenant Buchanan got underway from Ship Island for Biloxi. Buchanan commanded a task force made up of three steamers: the *Water Witch*, the *New London* and the *Henry Lewis*. His orders were to check on the report that a Rebel steamer was docked at Biloxi. Upon his arrival at Biloxi, he realized that the steamer had already departed, and he decided to demand the surrender of Biloxi instead. He informed city officials that a great destruction of property and life would occur if they did not surrender. A detachment of seamen and marines was sent to destroy the Confederate battery near the Biloxi Lighthouse. It was reported that two iron guns mounted on rough wooden carriages most likely made in Biloxi were removed. The largest was a 9-pounder, weighing 1,300 pounds, and the other a 6-pounder, weighing about 800 pounds. No ammunition was found, according to the report. In short order, Mayor James Fewell surrendered Biloxi. It was reported that Biloxi was almost deserted of a male population, but numerous women and children were present.

In 1865, the American Civil War ended, but it would take a very long time for the nation to recover. During the Civil War, one-third of the nation's population—roughly 970,000 individuals—was killed. The period immediately following the Civil War was a difficult time for the Mississippi Gulf Coast. Businesses and people struggled to make ends meet. Tourism was all but dead but would in due time be resurrected from the ashes of war. By the late 1800s, tourism along the Mississippi Gulf Coast began to grow again. Not only were new hotels being built, but boardinghouses and rental cottages were also available. Guests would again arrive by steamboat service from New Orleans and Mobile.

In Bay St. Louis, there were the Bay Saint Louis Hotel and the Clifton House. In Pass Christian, there were the Mexican Gulf Hotel, the New Magnolia Hotel, the Crescent Hotel and the Lynn Castle. Long Beach had the Brown Hotel. In Mississippi City, there were the Gulf View Hotel and Anniston Barnes Hotels. In Biloxi, there were the Nixon Hotel, the Shady Grove Hotel, the Magnolia Hotel, American Hotel, the Biloxi House and the Green Oaks Hotel. Ocean Springs had the Vancleave Hotel, the Illing House and the Ocean Springs Hotel. The Pascagoula area had the Pascagoula House and Dobson's Boarding House.

The arrival of the "iron horse" in 1869 would change the face of tourism. In time, travelers from the midwestern states and day travelers from New Orleans and Mobile would journey to the Mississippi Gulf Coast.

In 1886, the town covered an area of about 2,500 acres that ran along the shoreline. The bulk of the population was on the eastern extremity. The

View of Biloxi's Pass Christian Road in the late 1800s. It would be renamed Howard Avenue. *Courtesy of Alan Santa Cruz Collection.*

View of downtown Biloxi, late 1800s, looking west at the corner of Howard Avenue and Lameuse Street. *Courtesy of Alan Santa Cruz Collection.*

population was about 2,500, but during the summer months it went up to about 4,000. Visitors would come from southern cities and northern states to visit the Mississippi Gulf Coast. The Biloxi harbor was described as having a fleet of numerous small vessels, fishing smacks, lumber schooners, pleasure boats and yachts all lying at anchorage. In clear view from the shoreline is the harbor at Ship Island, where the masts of three-masted schooners and square-rigged vessels could be seen. Here, merchandise and other commodities were loaded and offloaded from smaller vessels.

Biloxi was a thriving cosmopolitan town with a wide variety of ethnicities. Its climate and proximity to the seashore made it a celebrated resort boasting several fine hotels. Some of the hotels were the Montrose, the Bossel, the Magnolia and the Fairview. The fishing from the rivers, the Bay of Biloxi and into the Mississippi Sound was described as excellent, with freshwater green trout (bass), speckled sea trout, red fish, sheepshead, Spanish mackerel and pompano being plentiful.

The best deer hunting was about ten miles north of Biloxi in the wooded areas. Duck and snipe were hunted in the Biloxi Bay with good results. Turkey hunting was conducted on the northern shores of Biloxi Bay.

From 1850 to 1881, the areas along Biloxi's shoreline and bay changed drastically. Still, we find that tourism and seafood continue to influence the growth and development of Biloxi. In 1878, a new industry began on the Mississippi Gulf Coast when a seafood cannery was established in Pascagoula. By 1881, Biloxi had four canning plants and seafood factories. These seafood factories canned fish, shrimp and oysters, as well as regional fruits and vegetables.

Chapter 4

Arrival of the Seafood Industry

The majority of people who live along our Gulf Coast, as well as tourists, have long enjoyed our local seafood. Seafood has been a mainstay of our historical past. The seafood of our coast attracted the Native Americans who would become the first harvesters of the Mississippi Sound. Even Iberville, as he explored our shores, could not resist tasting the oysters, which he indicated were rather good. As a young man, I worked in the factories and on shrimp boats. If you would have told me that the cannery row would one day be Casino Row, crawling with tourists, I would have laughed and made a sarcastic remark. The seafood industry of our recent past has faded into history. Yet it began with individuals who were not sure of themselves but were willing to take a chance.

One person we could thank would be Emperor Napoleon Bonaparte. In 1795, Napoleon offered 12,000 francs to anyone who could find a way to preserve food for his military forces. It wasn't until 1809 that Nicolas Appert, a French confectioner, won the prize. His discovery was that food adequately heated and sealed would be preserved. This process was the beginning of a new industry. In 1810, Englishman Peter Durand received a patent for using pottery, glass and tin-plated iron to can foods. The first cannery in America was established by Thomas Kensett in New York in 1812. This packing plant hermetically sealed cans of oysters, meats, fruits and vegetables. Then in 1818, Peter Durand presented his tin-plated can to America. In 1825, Thomas Kensett obtained an American patent for tin-plated cans.

The Barataria Canning Company on Point Cadet's Cannery Row (seen here in 1907) was built in the late 1800s. *Courtesy of Alan Santa Cruz Collection.*

The outbreak of the American Civil War inadvertently affected the Mississippi Gulf Coast. George W. Dunbar of New Orleans decided to send his sons to study in France for the duration of the war. Frank B. and George H. Dunbar would study fruit syrups and preserves. Upon their return, they started George H. Dunbar & Sons Cannery. In 1867, the cannery began to can shrimp and oysters. They developed a floating shrimp and oyster cannery off Grand Terre Island, Louisiana. One problem they encountered was that after coming in contact with the tin can, shrimp turned black due to the chemical reaction. The Dunbar brothers solved this by placing the shrimp in a cotton bag within the tin can. With the invention of ice-producing machinery, they moved their operation to New Orleans, and the new industry began to expand. The combination of train service, commercial ice production and these new canning methods would open the Mississippi Gulf Coast up to seafood canneries.

In 1878, a seafood cannery was established in Pascagoula, Mississippi. But the arrival of the railroad in 1880 opened the Gulf Coast seafood industry to broader markets. In 1881, Lopez, Elmer & Company opened its cannery near the foot of Reynoir Street on Back Bay Biloxi. The cannery owners were Laz Lopez, F. William Elmer, W.K.M. Dukate, William Gorenflo and James Maycock. These gentlemen invested $8,000. They possessed little or no knowledge of the industry but did possess a willingness to learn. Mr. Dukate had journeyed to Baltimore to purchase necessary equipment and to see how they canned oysters there. Oysters and shrimp were harvested only during certain times of the year, which created an off-season. During the off-season, the cannery would can locally grown figs. Little did these men realize the boom their small investment had begun.

In 1884, Laz Lopez and W.K.M. Dukate departed Lopez, Elmer & Company, which would change its name to Biloxi Canning Company. Lopez and Dukate united with Frank B. and George H. Dunbar to create the new company of Lopez, Dunbar's Sons & Company near Oak Street on the front beach. In 1893, this cannery introduced the steam oyster dredge to the Gulf Coast. This new innovation allowed oyster fishermen to harvest from reefs that had been unreachable by tonging. Biloxi Canning Company would have a few innovations as well. While other canneries packed their shrimp in cotton bags to prevent shrimp from turning black, Biloxi Canning Company used a wood veneer liner. The wooden liner absorbed moisture, leaving the shrimp palatable, and lessened breakage of the delicate shrimp.

In 1885, the industry began to boom with the addition of the Barataria Canning Company on East Beach. During the 1891 season, Biloxi Canning Company, Lopez, Dunbar's Sons & Company and Barataria Canning Company reported that 200,000 cases of oysters, 15,000 cases of shrimp and 5,000 cases of figs were packed, at a value of $350,000.

By 1892, four additional seafood canneries had been added in Biloxi. The Seacoast Packing Company, owned by Lopez, Dunbar's Sons & Company, operated on East Beach. William Gorenflo & Company, E.C. Joullian & Company and J.T. Maybury canneries were established on Back Bay. The seafood canneries were now employing about 1,200 people, with about $20,000 a week being spent on labor and raw materials. Laborers were paid about fifteen cents an hour. The oyster season began in September and ended in April. Shrimp were harvested in two seasons each year. The first season was February to March, and the second was September to October. During the off-season, figs were shipped to Biloxi from all along the Gulf Coast and purchased by the canneries at four cents per pound. By 1892, Biloxi had become the second-largest canner of oysters in the United States, with Baltimore in the number one spot.

In 1895, Biloxi's seafood canneries were shipping their product to many parts of the United States and Europe. Biloxi Canning Company was employing 350 laborers during the oyster season. One of its customers was a firm in England that had ordered 525 cases of canned shrimp. By 1896, Lopez, Dunbar's Sons & Company had become the largest oyster cannery outside of Baltimore. Its labor force consisted of about 600 individuals, a large fishing fleet, control of some large oyster reefs and a working relationship with independent fishermen. The 1898 Biloxi City Directory listed Biloxi

Biloxi schooners line factory piers. *Courtesy of Alan Santa Cruz Collection.*

Canning Company, William Gorenflo & Company and E.C. Joullian Packing Company on Back Bay, and Lopez Dunbar's Sons & Company and Barataria Canning Company on East Beach. Only one had electrical lighting. The others were using coal oil lamps or gas lighting. None of the canneries had heat, and all were using steam power to operate equipment.

In addition to an increase in seafood canneries, the industry's need for boats was met by an increase in large boatyards. Beginning with the first boats built in Biloxi from 1720 to the 1880s, skilled boat building had been practiced. As seafood canneries lined Biloxi's Back Bay and Point Cadet, so did boatyards. A common practice of placing them between factories was applied. Both large and small boatyards began to pepper the shores of Biloxi Bay and Point Cadet, including those owned by Bernard Taltavull, Martin Fountain and John Brander.

During the 1880s, large numbers of Slavic immigrants arrived to work in the seafood industry. The Slavs were from Dalmatia, one of the conquered provinces of the Austrian Empire. During this period, the population of Biloxi jumped from about 1,500 to about 3,000. In addition to canneries in Biloxi and Pascagoula, canneries began operating in Ocean Springs, North Biloxi, Mississippi City, Pass Christian and Bay St. Louis. Though Biloxi's population had doubled, there was still a shortage of laborers. In 1890, the first imported Bohemians, mostly of Polish descent, arrived from Baltimore. It wasn't long before all seafood canneries along the Gulf Coast were using these imported laborers from Baltimore. An agreement between the Gulf Coast canneries and the Baltimore canneries allowed these seasonal laborers to spend about nine months on the Gulf Coast and about three months in Baltimore.

In the Louisiana and Mississippi seafood industries, tokens were used to measure the work and at the end of the day to pay the workers. When a certain amount of peeled shrimp or opened oysters was reached, the worker received a token. A cup was designed to take these measurements. The token came in different values, such as one cup, five cups, ten cups and twenty cups. In Biloxi they were called shrimp nickels because the juices from the shrimp and oysters discolored them. At the end of the day, the tokens could be redeemed for cash. In the early years, they had to be redeemed at the factory store, which carried the basic needs of the family, from groceries to merchandise. This was the true meaning of saying "I owe my soul to the company store." The practice of using tokens would end when the federal government ruled that only the government could manufacture currency.

In 1900, the Biloxi canneries were spending about $250,000 for labor and raw materials. The canneries employed about 2,500 individuals with a weekly payroll of $25,000. These canneries were producing between 200,000 and 450,000 cases of oysters, 30,000 to 40,000 cases of shrimp and 10,000 to 15,000 cases of figs for worldwide distribution.

About 1900, Biloxi Canning Company was purchased by J.B. Humphreys and Charles Patten. Humphreys and Patten decided to operate the cannery

Biloxi schooners and luggers tied up on Point Cadet, Biloxi. These two styles of boats were designed by local master boat builders. *Courtesy of Alan Santa Cruz Collection.*

year round. They began canning string beans, sweet potatoes, okra, tomatoes, whole crabs and mullet, as well as figs, oysters and shrimp. In 1902, Lopez, Dunbar's Sons & Company became Lopez & Dukate Company and was considered the largest United States cannery packing oysters, shrimp and figs. By 1903, it had gained the title of largest company of its kind worldwide. This was due to good management, a large labor force, large privately cultivated oyster reefs, a fleet of more than sixty sail and steam vessels and a working relationship with independent fishermen. This cannery could steam up to four thousand barrels of oysters a day in four steam boxes.

By 1914, the Lopez & Dukate Company had become the Dunbar, Lopez & Dukate Canning Company. The 1914 Biloxi City Directory listed Biloxi Canning Company; Henry Diaz Fish & Oyster Packers; Biloxi Fish & Oyster Company; William Gorenflo Branch of Dunbar, Lopez & Dukate Canning Company; and Foster Fountain & Company on Biloxi Back Bay. Kennedy Lopez Oyster & Fish Company; Dunbar's, Lopez & Dukate Company; Lopez Greiner Fish & Oyster Company; Sea Food Canning & Packing Company; Gulf City Packing Company; George Terry & Sons Raw Oyster Company; Bourdon Castanera Packing Company; and Barataria Canning Company were located on East Beach, today's Casino Row.

The year 1914 also witnessed the beginning of World War I in Europe and the loss of imported laborers. When the United States entered World War I, the Bohemian laborers were not allowed to travel by rail from Baltimore to Biloxi. The fear was that one of them may sabotage the railroad because of their sympathy to Germany and its allies. The Barataria Company decided to send representatives to strawberry picking and trucking sections of Louisiana to recruit laborers. An article in the *Biloxi Daily Herald* on July 25, 1914, announced the arrival of these laborers from Louisiana. This would be the beginning of the migration of Cajun laborers to work the seafood industry. They arrived by train and were to live in the Barataria camps that were located on the Point. During the 1920s and 1930s, Biloxi factory owners and their representatives would continue traveling to southeast Louisiana to recruit laborers.

In the early 1900s, Slavic, Bohemian and Cajun immigrants were living on the Point and Back Bay. Most of the old fishing families were located on Back Bay and North Biloxi, in the present-day city of D'Iberville and St. Martin community. By 1915, the International Longshoremen's Association had started a branch of its union in Biloxi. The struggle between owners and fishermen was at an all-time high. In August, the fishermen voted to strike. Unfortunately, the economic effects of the war were being felt even in Biloxi.

A Biloxi lugger loaded with oysters, Point Cadet, Biloxi. *Courtesy of the Maritime & Seafood Industry Museum.*

The cannery owners refused to give in, and the struggle continued. Some unions stopped company boats, forcing the crews to turn back. When one schooner was set afire and a worker who was attacked by union men pulled a gun and seriously wounded one, the union requested that the Biloxi City Council intervene, but the owners refused to negotiate. Finally, the strike was broken, and on September 14 the workers returned to work. Some of the more defiant union members decided to form their own cannery. The Fishermen's Union Canning Company began operations and would last until 1918. In 1916, Biloxi Canning Company was bought by Dr. Babendreer. But the war had dealt the canneries a blow, and the days of the large canneries were few.

About 1912, the otter trawls first appeared in Mississippi waters. Over time, otter trawls would become the trawl net of Mississippi shrimpers. The otter trawl is a funnel-shaped net that narrows from the mouth to the bag, with two boards called doors. The two doors were positioned so that hydrodynamic force of the water opened the mouth of the net and kept it along the floor of the Mississippi Sound. A tickler chain weighs down the lower mouth of the net, causing the shrimp to jump out of the muddy bottom and into the net.

In 1923, Clarence Birdseye developed a method of quick freezing, and by 1930s it was being used to quick-freeze vegetables, fruit, seafood and meat. The seafood canning industry would continue but over time would see a reduction in canning until the last cannery closed.

By 1925, the larger seafood canneries had succumbed to the small seafood canneries. On East Beach, the Dunbar, Lopez & Dukate cannery, which was once the largest of its kind in the world, had been divided into four canneries. Those canneries were the Seafood Packing Company, DeJean, the Anticich Packing Company and the Deer Island Fish & Oyster Company. In addition to these seafood canneries, on East Beach there were Mavar Shrimp & Oyster Company Ltd., Quality Sea Food Inc., Dixie Fishery Oyster & Fish Packing, Wentzell Brothers Oyster Packers, Dubaz Brothers Seafood, Kuluz Brothers Packing Company Inc. and C.C. Company. Of the large seafood canneries, only the Biloxi Canning Company survived the economic crisis. Other seafood canneries on Back Bay were Latimer Packing, Gulf Company Seafood, Southern Sea Food Company, Southern Shell, Braun Company, Gulf Seafood Company, Gulf Fisheries Company and Foster Fountain Company.

In 1934, Bernard Taltavull, the owner of Bay View Packing Company, purchased Biloxi Canning Company and merged the two seafood canneries.

Biloxi schooners headed out to the fishing grounds. *Courtesy of Alan Santa Cruz Collection.*

Taltavull's Biloxi Canning Company had forty working boats, of which about two-thirds were used for oystering and the rest for shrimping. During the busy season, about 200 boatmen and 300 factory workers were employed. Taltavull had been in the cannery business for about seventeen years. He also owned a cannery in Bay St. Louis. Biloxi Canning Company was one of the last offering rent-free homes with water for workers who needed it. In 1934, Biloxi had about 500 boats with 1,500 boatmen and about 3,000 factory workers. In 1943, A.O. Soares and Roy Rosalis took over the operations of Biloxi Canning & Packing Company, but by 1956 it was nothing more than a seafood brokerage firm.

In September 1947, a hurricane struck the Mississippi Gulf Coast. It destroyed all the seafood factories on the Point and severely damaged those on Biloxi's Back Bay and North Shore.

The 1954 city directories indicated twenty-five seafood factories in Biloxi. On Back Bay, there were Aughinbaugh Canning Company, Biloxi Canning & Packing Company, Ray Canaan Company, Gollott & Canaan Seafood Company, Gollott & Knisey Seafood Company, Gollott Sea Food Company, Leckich & Fayard Sea Food Company, Southern Shell Fish Company, Suarez Sea Food, Taltavull B. Shrimp & Oyster Company and Weems Seafood Company. On Point Cadet's East Beach, there were Biloxi Seafood Packing Company, DeJean Packing Company, Dubaz Brothers Seafood Company, Gulf Central Seafood, Mavar Shrimp & Oyster Company, Moore's Seafood Company, Seacoast Packing Company, Victory Packing Company and Wentzell Brothers Seafood Company. In North Biloxi, there were two seafood factories, C.F. Gollott Seafood Company and R.T. Fournier & Son.

From 1950 to 1960, the seafood industry continued to struggle between good and bad years. Soon, another group of fishing families would call Biloxi their home.

On August 17, 1969, Hurricane Camille struck Mississippi, destroying the seafood factories and small-craft harbor on Biloxi's front beach, the Point and Back Bay.

Chapter 5

Pioneers of the Seafood Industry

James Maycock, Frederick William Elmer, Lazaro Lopez, William K.M. Dukate and William Gorenflo were the top five businessmen who changed the face of Biloxi. In 1880, these five men formed the first Biloxi seafood company, which was named Lopez, Elmer and Company. These five Biloxi gentlemen were from five totally different backgrounds.

Captain James Maycock

James Maycock was born in Hull, Yorkshire, England, in 1825. At the age of thirteen, he arrived in Biloxi in 1839 on an English ship. He had two things working against him from the beginning: he knew no one in Biloxi, and he was so young. In Biloxi, he showed that his character was one of integrity and he was a hard worker. Coming from a nation with a rich maritime history, he soon was involved in Biloxi's maritime industry. It was noted that he was a lake pilot as well as a captain.

While living in Biloxi, he met Mary Emily Carson. Mary Carson was born in Philadelphia around 1827. The Carson family migrated south in 1833. Mary Carson was six years old when the family arrived in Biloxi. James Maycock and Mary Carson were married in Biloxi on August 14, 1845. The Maycocks would have only one child, Elizabeth, who was born in 1850. Elizabeth would marry F. William Elmer on May 25, 1871.

Biloxi schooners rounded Point Comfort on Biloxi's Point Cadet. *Courtesy of Alan Santa Cruz Collection.*

During the Civil War, James Maycock served as a pilot for the United States Navy's Western Gulf Squadron under the command of Admiral David Farragut. After the Civil War, he returned to Biloxi and continued his maritime work along the Gulf Coast.

In 1880, James Maycock and Frederick William Elmer, his son-in-law, united with Lopez, Dukate and Gorenflo and formed the first seafood company in Biloxi. James Maycock was the oldest of the five at fifty-five years of age, while the other four were in their late twenties or early thirties.

Like the other members of the first seafood company, James Maycock was very active in his community. He was a member of the Masons and Biloxi's Volunteer Fire Company Number 1 and was one of the earliest and a very active member of the Episcopal Church of the Redeemer.

On March 19, 1892, James Maycock died suddenly at his residence on Main Street in Biloxi. On March 20, the funeral procession journeyed from his residence to the Church of the Redeemer and finally to the Biloxi Cemetery. The procession, consisting of Masons, members of the fire department and about five hundred carriages and citizens on foot, passed along the streets of Biloxi. The *Biloxi Daily Herald* reported, "Captain Maycock was consigned to the resting place by the side of the Mexican sea, where murmuring waves and gentle breezes will sing soft requiems over his grave till time on earth shall be no more."

On April 10, 1900, Mary Maycock died after a long illness and was laid to rest with her husband in the Biloxi Cemetery.

FREDERICK WILLIAM ELMER

Frederick Elmer was born on January 23, 1847, in Biloxi. His parents were Jacob Elmer of Switzerland and Barbara Gettendorf of Bavaria. He attended Spring Hill College in Mobile at the age of thirteen but left in 1863 at the age of sixteen to enlist in the Confederate army. He served in the cavalry in Company C of the Twenty-second Battalion. He was attached to General Wirt Adams's brigade and served under Captain John Wilkinson, Major Cal Roberts and Colonel Morman.

After the Civil War, Elmer returned to college and enrolled at Soule College in New Orleans. He returned to Biloxi and began teaching in Biloxi public schools. He became a principal within the school system and eventually served as the county superintendent of education. He married Elizabeth Maycock in Biloxi on May 25, 1871. From this union, ten children were born: Marie Ann, Ida, Nina, Cora, F.W. Elmer II, James, Inez, Margueryte, Edward and one more, unknown.

In 1880, Elmer, his father-in-law (Maycock), Lopez, Dukate and Gorenflo joined forces and formed the first seafood company in Biloxi. Elmer also became very active in politics at the city, county and state levels. He served on the Harrison County Board of Supervisors for many years. He was elected to the state legislature in 1883, 1892 and 1900. He also served as mayor of Biloxi in 1883 and 1892. He was selected by Mississippi Governor John Stone to act as Mississippi's pallbearer to former Confederate President Jefferson Davis, and he served as one of the organizers of the First National Bank of Biloxi.

On January 20, 1905, tragedy struck the Elmer family with the death of Elizabeth Maycock Elmer. Like all funerals of this period, she was laid out in her home. The funeral procession proceeded to the Church of the Redeemer for services and then to the Biloxi Cemetery for burial.

F. William Elmer never remarried but continued to be very active in the community. He was a member of the Masonic Lodge and the Port Commission and also served on the board of directors for the First National Bank. In the early 1920s, he became ill and suffered a heart attack. His illness continued until December 23, 1926, when he succumbed to a heart attack. Elmer's funeral procession began at his residence at 216 West Beach, which is the present-day parking lot of Mary Mahoney's Old French House Restaurant. Once again, Biloxians by the hundreds witnessed the funeral procession of a prominent citizen.

Lazaro Lopez Sr.

Lazaro Lopez Sr. was born on October 11, 1850, in Aviles, Spain. He received no formal education and very little education at home. In hopes of a new beginning, at only thirteen he traveled to Havana, Cuba, where his older brother Manuela, a merchant, lived. Here he would spend several years with his older brother. After leaving Cuba, Lopez traveled to the coast of Texas, and in 1868 he migrated to Biloxi.

When he arrived in the United States, he spoke very little English and had very little money. He worked various odd jobs until he opened a grocery business on Pass Christian Road, which would later become Howard Avenue. On August 19, 1871, three years after his arrival in Biloxi, Lazaro Lopez married Julia Dulion. Children born of this union were Lazaro, Arnaud, Julius, Juanita, Josephine, Clara, Theresa, Erena, Rowena and Noreta. Unfortunately, Juanita, Josephine and Clara would not live to reach adulthood. In 1880, Laz Lopez was among the organizers of the first seafood canning company on the Mississippi Gulf Coast, Lopez, Elmer & Company. In 1884, along with W.K.M. Dukate, Laz Lopez withdrew from the original company and joined the Dunbars of Louisiana to form Lopez, Dunbar's Sons and Company. The firm eventually grew to become the largest shrimp and oyster canning company in the world, including branch factories in Neptune, Violet and Rigolets, Louisiana.

Aside from his interest in the seafood business, Lazaro Lopez was active in many other enterprises. He operated L. Lopez and Company, which was a large mercantile store; held interest in Biloxi's two ice companies; and was a senior partner in T.P. Dulion & Company department store with his brother-in-law, T.P. Dulion. He built the first three-story brick building in Biloxi, located on the corner of Howard and Reynoir, which later became known as the W.V. Joyce Company. Lazaro Lopez was also an organizer and vice-president of the Bank of Biloxi and director of the Biloxi Electric Light Company.

In civic activities, Lazaro and Julia Lopez were exceedingly active. He served as an alderman-at-large for several years and was president of the Commercial Club from its organization until the time of his death. He was chairman of the committee formed to build the Church of the Nativity of the Blessed Virgin Mary after the fire of 1900 destroyed the original church building. The large bells that hang in the present Cathedral of the Nativity belfry were donated by the Lopez family. Lazaro Lopez was also a Masonic Knight Templar. After his death, Julia donated all five of the stained-glass windows in the sanctuary of Nativity as memorials to him in 1906.

Lazaro and Julia Lopez were some of the greatest benefactors of the Biloxi school system. The Lopez family donated both the grounds and buildings for two public schools. The Point Cadet Ward School, donated by the Lopez family in 1896, was located on Point Cadet near the Lopez-Dukate factory and labor tenements and was primarily for the education of the children of factory workers on the Point. In 1898, the Forest Park School (which was also known as West End School at one time) was donated by the Lopez family for the use of pupils below the third grade who lived west of Couevas Street. It was located on Porter Avenue at Irish Hill Drive. Years later, another elementary school was erected in his honor and was called the Lopez School. This memorial to Lazaro Lopez Sr., located on West Howard Avenue between Porter and Benachi, was closed in 1987 and torn down in 2000. The West End Elementary School at 300 St. John was renamed Lopez as a memorial to the man who made such a large contribution to the development of the public school system in Biloxi.

In the summer of 1903, Lazaro Lopez and some of his family toured Europe on a combination business and pleasure trip. After visiting his parents in Aviles, Spain, the family traveled to Rome. While in Rome, Lazaro Lopez was taken ill with kidney failure and died on September 25, 1903. The death of Lazaro Lopez shocked the entire city of Biloxi. He had earned the respect of every citizen, and the whole city would mourn him. Twenty-three committees were organized to handle his burial procession. On October 19, 1903, Lazaro Lopez's funeral procession marched though the city. Private citizens, business owners and civic and private organizations marched in the procession and lined the streets. It would be described as the largest funeral ever witnessed in the state of Mississippi. All schools in the city, whether public, private or parochial, were closed. All businesses were closed and draped in mourning. In the sad funeral procession were three hundred schoolchildren and two hundred members of the Oystermen's Protective Association. This tribute showed the tremendous respect and esteem that the citizens of Biloxi had for Lazaro Lopez Sr.

It is interesting to note the means by which Lazaro Lopez's body was returned to Biloxi. According to an article written in 1964 by Ray Thompson called "Know Your Coast," ships crossing the Atlantic in those days would not carry a corpse. However, Lazaro Lopez had been so emphatic in his wish to be buried in his beloved Biloxi that the family was forced to find a way around this policy. Consequently, the family was able to ship the body home only by listing it on the cargo manifest as "a piece of Italian marble statuary." Lazaro Lopez's wife, Julia Dulion, peacefully passed away on June 30, 1918.

William K.M. Dukate

William Dukate was born in Fredericksburg, Washington County, Indiana, on August 10, 1852. His father was John S. Dukate. He gained employment at the Louisville & Nashville Railroad, where he would be employed as a telegraph operator. In 1875, his work as a telegraph operator brought him to Biloxi. He would continue in their employment as an L&N Railroad agent from 1879 until 1883. In Biloxi, he met Linda Rose Lienhard, the daughter of Peter Lienhard and Malinda Seaman. A romance ensued, and on April 27, 1878, William Dukate and Linda Lienhard were married.

The year 1880 proved to be a banner one for William Dukate, who joined Lazaro Lopez, William Gorenflo, James Maycock and William Elmer to form Lopez, Elmer & Company. Since none of these gentlemen knew anything about this industry, William Dukate was selected to go to Baltimore and learn. In Baltimore's factories, William Dukate purchased the necessary machinery and other equipment for coving oysters and canning shrimp.

William and Linda Dukate had their first child, Elbert L. Dukate, in 1881 and their second, Eula Dukate, in 1883. Their family would continue to grow with the births of Vera in 1886, Leola in 1888, Irma in 1890 and Beula in 1900. The Dukate family home was built by Theodore Brune of Stuttgart, Germany. Brune had studied architecture in German universities

Mountains of oyster shells piled so high they conceal the factory. *Courtesy of Alan Santa Cruz Collection.*

and graduated with high honors. He came to America and worked in New York and St. Louis before moving to New Orleans. Some of his noted works in Biloxi were the Dukate Theatre, Biloxi Yacht Club and Nativity BVM Church. The Dukate family home was a Queen Anne style, which was described as a "dream in an architectural sense." Unfortunately, like most of the large homes of this period, the Dukate home has long faded into history.

By 1884, Lazaro Lopez and William Dukate had withdrawn from Lopez, Elmer & Company to start another canning factory under the name of Lopez, Dunbar & Sons Company. The next year, they organized the Barataria Canning Company with businessmen from New Orleans. Originally there were two factories—one in New Orleans and one in Biloxi. In the 1880s, Biloxi was the only place in the United States—with the exception of New Orleans—where shrimp was canned. The firm of Lopez and Dukate was formed and became the largest individual packer of oysters, shrimp and figs in the United States. On May 1, 1901, 525 rail cars of canned oysters and shrimp were shipped from their factories. In 1901, Lopez and Dukate owned sixty-odd vessels under both sail and steam. In 1900, Laz Lopez was listed as head of the firm and was supervising the company's large interests, and William Dukate was in charge of the sales department. Both Laz Lopez and William Dukate are considered to have been some of the most progressive men in Mississippi.

William Dukate was also very active in other business enterprises. Along the Gulf Coast, he was president of the First National Bank of Biloxi, secretary of the Biloxi Artesian Ice Manufacturing Company and a director of the Gulf Coast Life Insurance Company of Beauvoir and T.J. Rosell Manufacturing. He was also involved with other sections of our state. He was a director in the Columbia Ice and Power Company and Pearl Realty Company of Columbia, Mississippi.

In Biloxi, he was considered a public-spirited individual. In civic activities and charities, William Dukate was exceedingly active. At the time of his death, he was a member of the Biloxi Board of School Trustees and had been for thirty years. He and his partners were benefactors of the Biloxi schools. In 1898, William K.M. Dukate donated the Back Bay School, William F. Gorenflo donated the property for the Back Bay School and Lazaro and Julia Lopez donated the West End School and the property on the corner of Porter Avenue and Cemetery Street (present-day Irish Hill Drive). Dukate always took an interest in organizations that promoted projects for the good of the community. He was also very active in the Biloxi Commercial Club, today's Biloxi Chamber of Commerce. He belonged to Biloxi Lodge No. 606 BPO Elks and the Biloxi Magnolia Lodge #120 of Free and Accepted Masons.

In 1916, it appears that William Dukate had not been in the best of health, and on March 29, 1916, he passed away. He was laid to rest on March 31 in the Biloxi Cemetery. The Reverend H.F. Tolle, pastor of the Methodist Episcopal Church of Biloxi, officiated at the services. All Biloxi businesses and schools were closed in his honor. Hundreds of Biloxians and schoolchildren lined Howard Avenue to view the funeral procession, which consisted of automobiles, carriages and members of the Masons and Elks marching to honor William K.M. Dukate. Prominent men of the seafood industry also attending included William Gorenflo, Frank H. Gorenflo, Sol Brown, James Dunbar, George Dunbar and Dr. Jestremski. These gentlemen represented the 1916 seafood industry in Louisiana, Mississippi and Alabama. Twenty-three years later, on October 18, 1939, Dukate's wife, Linda Rose Lienhard Dukate, passed away.

In 1924, the Biloxi Public School Administration dedicated Dukate Elementary School on East Howard Avenue in honor of William K.M. Dukate. Longtime Biloxians remember this school and still refer to the curve in Howard Avenue as "Dukate Curve." Sometime in the early 1970s, Dukate Elementary was moved to the Old Biloxi High School on Howard Avenue. In 2003, Dukate Elementary School was closed for good. After the closing of Dukate Elementary, the Biloxi Public School Annex Building on Father Ryan was eventually renamed the Dukate Building.

William F. Gorenflo

William F. Gorenflo was born in Bay St. Louis, Mississippi, on June 5, 1852. When he was eight or nine years old, the family moved from Bay St. Louis to Biloxi. During the Civil War, William Gorenflo enlisted and joined Company E, Third Mississippi Regiment. After several battles, he was captured at Vicksburg and became a prisoner of war at Fort Morgan, Alabama. He escaped from Fort Morgan and returned to Biloxi. In Biloxi, he met and began courting Theresa Llado. On April 9, 1874, William Gorenflo and Theresa Llado were married in Biloxi. From this union, eight children would be born: William, Frank, Henry, Daniel, Alfred, Aline, Ernestine and Nora.

In 1880, with the birth of Lopez, Elmer & Company, Biloxi's first seafood factory, William Gorenflo launched a career in the seafood industry. Double tragedy struck the Gorenflo family in 1892. On January 24, William's beloved Theresa died during childbirth, and on July 9 their infant son Alfred died. That same year, William Gorenflo & Company was formed and located on

Factory and schooners on Biloxi's Point Cadet. *Courtesy of Alan Santa Cruz Collection.*

Biloxi's Back Bay. Between 1900 and 1913, Gorenflo would open factories in Houma, Louisiana, and Fernandina, Florida. In 1914, he opened a factory in Brunswick, Georgia. By 1914, William Gorenflo & Company had evolved into the William Gorenflo Branch of Dunbar, Lopez & Dukate. In addition to the seafood industry, Gorenflo helped organize the First National Bank of Biloxi. He would serve as its first vice-president. He also was active in Biloxi politics, serving as an alderman.

William Gorenflo's philanthropies helped the Biloxi public and parochial schools, many charities and civic organizations. In 1898, William Gorenflo donated the property for the Back Bay School on Main Street. In 1924, a new Back Bay School was constructed on Lameuse Street. On September 5, 1924, the Biloxi Public School's Board of Trustees officially named the new school Gorenflo in honor of William Gorenflo. For several years, the old Back Bay School site remained vacant. William Gorenflo had not deeded the property to the city, so he donated it to the Sisters of Mercy. This would become the site of St. John's Parochial School.

William Gorenflo was active in the Biloxi Lodge No. 606 BPO Elks. He was a life member and served on the building committee that erected Biloxi's Elk Lodge. He was also a life member of the Biloxi Masonic Lodge and became a thirty-second-degree Mason and Shriner.

William F. Gorenflo passed away on February 17, 1932. Once again, hundreds lined the funeral route. The funeral procession started at his daughter's home on Main Street and went to Bayview Avenue west to Lameuse and then south on Lameuse to the Church of the Redeemer. At Gorenflo and St. John's Schools, the schoolchildren, with heads bowed, witnessed the funeral procession. Reverend E.A. DeMiller officiated over the services. The Biloxi Elks and Biloxi Masons conducted services for their lodges. The funeral procession continued to the Biloxi Cemetery, where William Gorenflo was laid to rest.

As pioneers of the Biloxi seafood industry, Captain James Maycock, Frederick William Elmer, Lazaro Lopez, William K.M. Dukate and William Gorenflo left a historical legacy that continues to this day.

Chapter 6

Slavonian, Bohemian, Cajun and Vietnamese Seafood Workers

From 1880 to present times, a diverse group of seafood industry workers have supplied laborers for Biloxi's seafood industry. These individuals were of French, Spanish, Italian, Irish, African and other ethnic groups. The main work force would be made of Slavonians, Bohemians, Cajuns and Vietnamese workers who arrived at different times.

In the early years of the industry, Slavonian, Bohemian and Cajun men, women and children worked long hours in the factories. The use of children in the United States labor force did not go unnoticed. In the early 1900s, the National Child Labor Committee was formed. By 1904, they were leaders of an aggressive campaign to create child labor reform. Their goal was reached in 1938 when the Fair Labor Standards Act was passed. This act had child labor provisions that ended the practice of using children in the labor force.

The Slavonians, Bohemians, Cajuns and Vietnamese all had the same goal of making a better life for themselves and their children. Each would try to hold on to their culture while becoming productive citizens. Their values would be passed from generation to generation. Their desire was for their children to receive a good education that would allow them to enter other professions and therefore have an easier life than they had.

Slavonians

During the mid-1800s, a large number of Slavic immigrants arrived in Louisiana and Mississippi seeking work in the fishing industry. In 1880, Biloxi had opened a new seafood cannery, and in due time, these Slavonians began to seek work in various other Biloxi canneries. During this period, the population of Biloxi jumped from about 1,500 to about 3,000. In many cases, the Slavonian men arrived first and worked hard to save money to send for their families. These Slavonian immigrants were from the Dalmatia Coast that in 1822 had been conquered by the Austrian-Hungarian Empire. At the end of World War I, Dalmatia became part of Yugoslavia, and many of Biloxi's Slavonians began to refer to themselves as Yugoslavian. In the 1990s, more conflict arose in Yugoslavia and resulted in Dalmatia becoming part of present-day Croatia.

As the Slavonian population of Biloxi grew, they became the mainstay of the seafood industry. They also faced prejudices due to their language, dress and cultural customs. Their work ethics would in time earn them respect

Dedication of the Slavonian Lodge, 1938. *Left to right*: Alonzo Gabrich, Vincent Kuluz, Steve Kuljis, John Mavar Sr., Tony Pitalo, Luka Kuljis and Matre Pitalo. *Courtesy of the Slavonian Lodge.*

in the community. The men and older boys worked the boats while the women and children worked in the seafood canneries. They would witness the growth and ever-expanding seafood industry in Biloxi. By 1900, a large community of Slavic immigrants had settled around the canning plants on Point Cadet and Back Bay. In the early years, wages were low, and there were times when no work was available. During these hard times, some would journey to barrier islands to collect driftwood to be used for firewood. If someone in the Slavonian community could not work or died, all others would work together to help the family. This was the inspiration that led to the founding of the Slavonian Association.

On December 18, 1913, the *Biloxi Daily Herald* reported that the Slavonian Benevolent Association of St. Nicola had formed and was working on its incorporation. St. Nicola would be the patron saint of the association. The charter members were Philip Williams, Jacob Stanovich, Victor Simonich, Ivan Letich, Blas Mijan, Nikola Skrmetta, Ivan Lovetich, Tony Rosetti, Marko Mulat and Ivan Gabrich. The newly elected officers were Jacob Stanovich, president; Nick Skrmetta, vice-president; Victor Simonich, treasurer; Philip Williams, secretary; Ivan Budinich, outer guard; Ivan Gabrich, inner guard. The new organization felt that the membership would grow to two hundred.

Dues were set at $1 a month. One of its main purposes was to provide sick and death benefits for the membership. If there was a family death, the member was entitled to $150 and, in the case of illness, $10 a month for ten months. During the early years, sickness and deaths consumed the association's funds. Between 1918 and 1957, the finances of the association were ample enough to meet those needs. Another reason for the creation of the association was to teach their Slavonian culture to their children, help members become productive citizens and to have social gatherings. Like many immigrants to America, the members wanted more for their children. They knew the importance of a good education, which was stressed and enforced by each family.

The first state charter listed the lodge as the Austrian Benevolent Association. In 1918, the lodge under President Steve Sekul amended the charter and changed it to Siovinsko Dobrotverno Drustvo, or Slavonian Benevolent Association. Around 1954, the dues were raised to $1.50 a month. The Slavonian Lodge, as it is known locally, sponsored twice-a-year suppers for members and their wives and an annual bazaar and was part of the Blessing of the Fleet celebrations.

Despite the Depression, in 1930 the lodge was able to build a new hall for $4,000. The new lodge was built at 1624 East Howard Avenue between

Cedar and Pine Streets. By the late 1930s, the members had outgrown this building. Land was purchased on the corner of Myrtle and First Streets, and in 1938, the new lodge was constructed for $25,000. The president was Steve Kuljis; vice-president, Vincent Kuluz; and secretary-treasurer, Tony M. Pitalo. Mr. Pitalo would serve in this position for twenty-two years. Between 1938 and 1957, the building was air conditioned and a new addition was constructed. The membership in 1957 was 136 men. The officers were President Jake Mladinich, Vice-president Dr. Steve Pitalo and Secretary-treasurer Steve Thian.

In the early years, for a male to join the Slavonian Lodge, his mother and father had to be of Slavonian descent. In time, the rules were changed to allow men whose fathers were Slavonian. Today, any male whose father or mother is Slavonian may apply for membership.

The Slavonian women are also now involved in the lodge through their own organization, called the Ladies Auxiliary. On June 3, 1953, it was founded for the purpose of providing an organization for social, charitable and civic activities. The Slavonian Ladies Auxiliary was chartered with the State of Mississippi on June 28, 1953. The officers were Miss Mary Kersanac, president; Mrs. Ruby Rosetti, vice-president; Mrs. Anastasia

The Slavonian Lodge was destroyed by Hurricane Katrina 2005, and a new Slavonian Lodge was built in 2010. *Photo by Mike Skrnich.*

Jurich, treasurer; and Mrs. Margaret Soper, secretary. For many years, the women have been involved side by side with the men helping the lodge achieve many goals. Now they have their own organization.

To promote and provide for the general welfare of its members, the ladies' biggest fundraiser is the Christmas pusharata sale. Pusharatas are basically a Slavonian doughnut with a kick. The delicacies are filled with the right combination of fruits, cinnamon, nutmeg and raisins. They are then covered with a warm glaze made from powdered sugar, evaporated milk and almond extract. During this fundraiser, over 1,300 dozen pusharatas are sold to local coast residents.

The Slavonian Lodge became the culture center for the Slavonians. In the lodge and on its grounds, wedding receptions, banquets, dinners and fundraisers were held. The walls were decorated with large photographs of Biloxi schooners. Over time, many Slavonians would become owners of seafood factories. Today in addition to the seafood industry, many have entered professional fields such as law, medicine and politics.

In 2005, the Slavonian Lodge on the corner of Myrtle and First Streets was destroyed by Hurricane Katrina. Immediately, plans were made to construct a new lodge on the corner of Maple and Third Streets. The new lodge was completed in March 2010 with hopes of not only bringing home its old members but also welcoming more of the younger men of Slavonian descent. The rebuilding of the Slavonian Lodge was important to the recovery of Biloxi. The culture of the Slavonian people became part of the fabric that makes Biloxi such a diverse and special place to live and work.

BOHEMIANS

Due to the fact that Baltimore was known for its seafood canning factories, it developed strong commercial ties to Biloxi beginning around 1880. As the industry grew in Biloxi, the need for labor also grew. A lot of immigrant labor was available in Baltimore during its off-season. Around 1890, representatives of the Biloxi canneries and Baltimore canneries came to an agreement that would benefit both. The Bohemian seafood workers would journey by train from Baltimore and work the shrimp and oyster season for nine months in Biloxi. Then they would journey back to Baltimore and work there for three months.

Biloxi and Baltimore factory owners built seafood workers' camps. The Bohemian seafood workers were housed in these cannery-owned buildings. The best way to describe them was row houses. They were long, single structures with partitioned apartments for each family. Referred to as Bohemians, these workers were mostly Polish with some Germans. On January 11, 1890, the *Biloxi Herald* reported, "Locals stared in amazement at these strangers garbed in peasant dress."

The whole family worked; while the women and children were in the factories, the men and older boys were on the boats. Like the Slavonians, they practiced their culture not only in their dress but also in their style of music and dancing. The annual immigration of Bohemian seafood workers would last twenty-eight years. The federal census of 1900 recorded Bohemian families in the seafood camps. Some of the families were listed as being from Germany, Prussia and Poland. Some of the family names—some of which were most likely misspelled by the census workers—were Jablinski, Scofrinsky, Hneziky, Miliski, Lerandusky, Sarsada, Wasnass, Pawlok, Boitkusky, Vrankosky, Antufsky and Cakuta. The majority of their children were listed as born in Maryland. During the course of these twenty-eight years, some workers decided to stay in Biloxi. The majority would return to Baltimore, bringing an end to the Biloxi-Baltimore era.

Cajuns

The Cajuns were first recruited to work the seafood industry on the Mississippi Gulf Coast in 1914 after factory agents went to Louisiana. The first Cajun laborers arrived by train on July 25, 1914. After their arrival, they were taken to the Barataria camps that were located on Point Cadet. During the 1920s and 1930s, Biloxi factory owners and their representatives would travel to southeast Louisiana to recruit laborers from places like Broussard, Youngsville, New Iberia, Jeanerette, Erath and St. Martinville. The surnames of some of these early families were Borel, Boudreaux, Breaux, Broussard, Comeaux, Delahoussaye, Duet, Gautreaux, Martin, Olier, Thibodeaux, Theriot, Viator and Use.

Like the Slavonians and Bohemians before them, the Cajuns faced prejudices because of their culture and the language barrier. There was a fear among existing workers of losing jobs because of these new seafood workers. Not only local labor but also some Slavonian and Bohemian seafood workers felt that their jobs were being threatened. Even though these Louisiana natives were United States citizens, they were told to go back to their own country. Only time would dispel the rumors of lost jobs. The hardworking

Cajuns eventually earned the respect of the other seafood workers. In return, they would respect and become friends with the Slavonian and Bohemian seafood workers.

The Cajuns brought many cultural customs with them, including the old custom of having neighborhood dances, or "faire do-do." From 1914 to the 1930s, their numbers continued to increase as family and friends in Louisiana joined them. Soon there was talk of a French club, which went from talk to meetings in 1933. French fishermen wanted not only a social club but also an organization to help members when the need was there. Some of these meetings were held at the Community House on Back Bay. Individuals of French descent from North Biloxi, Back Bay and the Point were in attendance.

These French fishermen were proud of both their heritage and their profession. They desired better lives for themselves and a good education for their children. They wanted to become productive citizens and retain their cultural identity at the same time. The purpose of the society was to create a means to hold on to and teach their French culture, to educate its members and help them become productive citizens, to assist individuals of

Funeral of a member of the Fleur de Lis Society. *Courtesy of the Maritime & Seafood Industry Museum.*

their community who were in need and to have social gatherings or faire do-dos with friends and family.

William Estopinal, a lawyer of French descent and attorney for the fishermen's union, was approached to help them organize the society. Another individual who aided the organizers was Roy Hepler, secretary-treasurer of the fishermen's union, who suggested the name Fleur de Lis. The fleur de lis (or flower of lily of France) was a longtime symbol of France and coat of arms of the House of Bourbon, which ruled France from 1589 to 1793. The name was adopted, and the Fleur de Lis Society was born.

On February 17, 1934, the charter of incorporation of the Fleur de Lis Society was filed in Harrison County. George B. Wink, justice of the peace, signed the charter and forwarded it to the state capital in Jackson, Mississippi, for approval. Walker Wood, secretary of state, accepted the ten-dollar fee and referred it to Greek L. Rice, the attorney general, for his opinion. The Fleur de Lis Society charter of incorporation was approved March 25, 1934. Those who signed the charter were Emar J. Boudreaux, P.A. Songe Sr. and Eli Romero, all from Biloxi; Homer Barras from North Biloxi; and William Estopinal from Gulfport. P.A. Songe was elected the Fleur de Lis Society's first president.

Some of the early members were Lucian Arcenaux, August Bairllaux, Adam Boudreaux, Clarence Boudwin, Xavier Boudwin, Alcee Broussard, Arthur Broussard, Jake Broussard, Lesson Broussard, Noah Broussard, Laurent Duplain, Ben Hebert (who was elected first secretary), James Higginbotham Sr., Clomer Leaux, Seagus Leaux, Horace Mouton, Lucian Olier, Adolph Romero, Oliver Romero, Fernando Smith, Laurent Smith, August Theriot and Wallace Vallo. The original charter membership was about seventy-five individuals.

The early membership laws of the Fleur de Lis Society were very strict. To join, one had to prove his French ancestry three generations back and speak French. These early membership meetings were all conducted in French. Over time, the children and grandchildren of local French families were not taught the French language at home. In 1952, the society dropped the requirement of strictly French-speaking individuals and began to conduct meetings in English. Membership eligibility still required French ancestry back three generations.

Early meetings were held in private homes and at the old St. Michael's School that was located on First Street between Myrtle and Cedar Streets. After numerous contributions and fundraising efforts, the society purchased some property and a house owned by S.P. Lee Sr. on the corner of East Howard Avenue and Seymour Lane. This is its present-day location.

After its destruction in Hurricane Katrina, a new Fleur de Lis Society building was built in the original location.

About 1953, the Fleur de Lis Ladies Auxiliary was formed by ten ladies, who elected Mary Duplain its first president. Together, the society and the Ladies Auxiliary began to host various functions that have become Biloxi traditions. By the late 1950s, the regular paid-up membership numbered about 150. In 1957, James Higginbotham was president, Norman Broussard was vice-president, Adam Boudreaux was treasurer and Ralston Primeaux was secretary, with a twenty-one-man board of directors.

In 2005, the building was destroyed by Hurricane Katrina and rebuilt in no time. Before Hurricane Katrina, the Fleur de Lis Society and Ladies Auxiliary memberships were up to about 350 to 400. Today, the men and women's membership is about 300 and climbing. Male membership eligibility still requires French ancestry on either the mother's or father's side, while members of the Ladies Auxiliary can be of French descent or contingent on their husbands' membership in the society. The society also has some honorary members on its membership rolls. The return of the Fleur de Lis Society in Biloxi was a sign of historical and cultural recovery in Biloxi. The Cajun culture, like those of the Slavonians and Bohemians, added to the makeup of Biloxi's cultural heritage.

VIETNAMESE

In 1967, the Vietnamese began arriving in Biloxi, and in no time their community began to flourish. Just like the Slavonians, Bohemians and Cajuns before them, the Vietnamese would be drawn into the seafood industry. It sounded like a page out of the Biloxi early seafood industry history: the Vietnamese men worked the boats and the women and children worked in the factories. Vietnamese families would locate near the seafood factories of Back Bay and the Point. They were seeking a better life and education for their children. At the same time, like the Slavonians and Cajuns, they wanted to hold on to their culture. The strong and close-knit Vietnamese families worked hard and were thrifty.

Like the Slavonians, Bohemians and Cajuns had encountered before, language and cultural differences were a difficult barrier to transcend. In addition to these barriers, race was a factor as well. In the past, with the arrival of new seafood workers, the previous workers felt threatened and feared these new workers would soon take their jobs. This friction was most apparent with the American boatmen toward the Vietnamese boatmen. Of course, not all of the American boatmen were opposed to the Vietnamese boatmen. In fact, some attempted to help them get started.

One of the major problems was that the seafood industry already had a strained field of work. The additional Vietnamese boats put more pressure on all of the boatmen. Another problem had to do with shrimping and oystering techniques. The Vietnamese used what was called "the chopstick method," which was two poles extending from the bow of their boat to push the trawls through the sound. The American boatmen pulled their trawls. While the Vietnamese worked the sound north to south, the American boatmen worked east to west.

Add to the mix the cultural and language differences, and Biloxi was sitting on a time bomb. Finally, a committee composed of Vietnamese and American fishermen, city officials and marine officials was formed. This would dispel rumors, furnish a translated version of Coast Guard rules and explain traditional fishing practices. This was the beginning of an understanding between the Vietnamese and American fishermen. Time has ushered in an era of understanding and acceptance.

Like the Slavonians, Bohemians and Cajuns before them, the Vietnamese have strong ties to the Catholic Church. Many are parishioners of various Catholic churches along the Mississippi Gulf Coast. As in the past, the church is the center of social and culture affairs for the Vietnamese community. On

Church of the Vietnamese Martyrs, Biloxi.

June 26, 2003, the Church of the Vietnamese Martyrs was dedicated on Oak Street in Biloxi. The services are conducted in Vietnamese, and the community conducts many cultural events. The Vietnamese have learned to adjust to American culture while still holding on to their culture and values.

The Slavonians, Bohemians, Cajuns and Vietnamese each had to learn to live and work together. Culture and language did not stop the interrelationship between groups. One of the common denominators was that all of them started with little or nothing. Over time, each group has become culturally more mainstream American. The successes of each group in business and professional fields have marked their place in Biloxi history. The Slavonians, Bohemians, Cajuns and Vietnamese, like so many explorers, settlers and immigrants, have added to the rich cultural heritage of Biloxi and the Mississippi Gulf Coast.

Chapter 7

Wooden Boat Races and White-Winged Queens

Historically, regattas have occurred along our shores since at least 1849, when New Orleans papers were abuzz with the Biloxi Regatta. In the 1849 regatta, there were eighteen boats in two classes. The first-class winner, *Undine* of Mobile, was awarded a Silver Pitcher, and second place went to the *Mary Ana*, which was awarded a Silver Goblet. In the second-class sailboats, the *Mary Bella*, owned by H.T. Sherman of New Orleans, received a Silver Cup. There was also a two-and-a-half-mile rowing match between four skiffs with two men per skiff, but unfortunately the winner was not recorded.

Most of the regattas occurred during August and September. During August 1888, the *Biloxi Herald* reported the Biloxi Regatta as a "Gala Day." The regatta sported fifteen entries in three classes. The judges and timekeepers were mostly New Orleans natives and included an ex-sheriff, former mayor and the chief of the New Orleans Fire Department. The classes were schooners, sloops and cat-rigged boats. The schooner class winner was the *H.T. Howard*, entered by Mr. Harvey and sailed by Mr. De Loch. The *H.T. Howard* finished the course twenty minutes and twelve seconds ahead of the Mr. Fontaine boat *Italia*. The sloop *Susie B.* was the winner in the sloop class. The *Winonah*, entered by Frank Raymond, won the cat-rigged class. About five hundred coast natives filled the piers and boats to watch the regatta.

By the turn of the century, the Mississippi Sound had witnessed and hosted various regattas. On August 4, 1900, the *Biloxi Herald* reported that the "last big regatta of the summer" would be held at Biloxi. Because Biloxi

A Biloxi schooner sails past the Biloxi Yacht Club during the annual regatta. *Courtesy of Alan Santa Cruz Collection.*

was midway between New Orleans and Mobile, a good turnout of boats was expected. The three-day event was scheduled with catboats and open sloops on Friday, August 17. The cabin sloops and schooner events were on Saturday, August 18, and the flat-bottomed schooners were on Sunday, August 19. The local watering holes hosted dances each night of the regatta.

On the first day of the regatta, there were thirty-four entries in the catboat and open sloop classes. The winners in each respective class were *Nana*, *Belta S.*, *El Heirie*, *Flirt*, *Emma Johnson*, *Amazon* and *Royal Flush*. On the second day, the *Queen of the Fleet* won for schooners forty feet and over. The *Mystery* took the schooners under forty feet, while in two classes of cabin sloops *Annie* and *Susie B.* won unopposed. The *Nydia* won the cabin sloop class under twenty-five feet. In the cabin sloop under twenty-one feet, the *Chrutti* took its class. On Sunday, the *Thelma* took the barge race while the *Alida* took the catboat race. This ended one of the most successful regattas in Biloxi.

On July 4, 1901, everything was in readiness for the firing of the first gun of that year's regatta. Immediately following the races, the Montross Hotel, later known as the Riviera, would host a "Grand Dance." A second dance was to be hosted by the Montross on Saturday night. In addition to the races and dances, the Catholic Women's Aide Society was hosting a fair to benefit the church building fund. The fair was held at the residence of Mrs. Bachino on the front beach just east of the Montross Hotel. Additionally, the aide's society sold refreshments to dance patrons from a beautifully decorated booth on the Montross Lawn. The *Biloxi Herald* reported that many danced and took in the fair, as well as enjoying the luncheon prepared by the aide's society.

The first day's races were declared a great success, with an immense crowd lining the beaches and piers of Biloxi. In the open sloops eighteen feet and under twenty-three categories, the *Irma*, owned by R. Gautier, took first place. The *Imp* took first place in the twenty-three feet and over, while the *El Heire* took the open sloop, half rater category. The open cabin sloop *The Flirt* sailed the course but was not an entry.

The second day saw a "Big Fleet of Cats"—catboats—with twenty-two boats entered in the contest. The *Bessie B.* took first place in the launch race, while P. Harvey's *Fanchon* took the first-class catboats. In the second class, J. Lopez's *Noretta L.* was the winner, with C. Bellande's *Davis Eros* taking the third class. In the fourth class, U. Desporte's *Alide T.* won.

On the last day of racing, a strong wind caused problems for some of the contestants. Seven of the boats quit the races, with two becoming disabled. In the cabin sloops over thirty-three feet, the *Annie*, owned by T.G. Bush, was the winner. In the cabin sloops twenty-five feet and under thirty-three, the *Juno*, owned by Joe Pal, was the winner. *The Flirt* capsized during the event and was assisted by the *Charles W. Mackie*, which also became disabled. In the schooners over forty feet category, *The American Girl*, owned by John Trochesset, was the winner.

On July 9, 1901, the *Biloxi Herald* reported that the regatta was very successful and that all resources should be used to make it an annual event. The Fourth of July regattas and races of the white-winged schooners would become an annual tradition.

Schooner races most likely occurred as friendly competition as fishing schooners were heading out to fishing grounds or returning to the factory. Of course, the first boat to the oyster reef or shrimping grounds gets the prime location. Also if you get to the factory pier first, your catch would be unloaded first. Soon this pastime caught on, with the factory owner boasting who had the fastest schooner. By 1900, it became part of the Fourth of July regattas as the Gulf Fishermen's Schooner Championship.

The *Biloxi Daily Herald* of July 8, 1929, heralded that the upcoming Gulf Fishermen's Schooner Championship Classic was going to take place during the Biloxi Yacht Club Thirtieth Regatta. The schooners listed in the race were the *I. Heidenheim, Julia Delacruz, H.E. Gumbel, Mary Foster* and a new schooner called the *Mary Margaret.*

The schooners had been out of the water for several days to allow the vessels to dry out. Their bottoms had been stripped and a new coat of paint

Biloxi schooner *H.E. Gumbel* sails past the Biloxi Yacht Club. *Courtesy of the Maritime & Seafood Industry Museum.*

applied. These labors were all done to make the schooners lighter and faster. The *Daily Herald* reported that the crews of the "white beauties of the Gulf water" had been working at a feverous pitch.

These races that pitted factory against factory, owners against owners, and fishermen against fishermen created a spirit of rivalry not only in the participants but in their families and locals as well. As the *Daily Herald* reported, the enthusiasm had been worked up to its highest pitch.

Captain Martin Fountain—referred to as the dean of Biloxi skippers—stated that they would do their best by the boats and get the best results from them. Captain Fountain skippered the *H.E. Gumbel* of the Seafood Packing Company. Captain Fountain indicated that all the boats were in first-class condition and that any one of them could be the winner. When he was asked about Biloxi Canning Company's new entry, the *Mary Margaret*, he replied: "She's a mighty fine boat, fast and dependable but we cannot tell until after the race about her racing ability."

Biloxi schooner *Mary Margaret* with wind filling its sails during the annual regatta. *Courtesy of the Maritime & Seafood Industry Museum.*

The defending champion of the 1928 race was Captain Amos Ross and the schooner *I. Heidenheim*. Captain Ross indicated that he knew the *I. Heidenheim* and would get the best results from it. Captain Ross stated: "She is well adapted to sail under any weather condition whether light breeze, heavy winds, or choppy water."

The *Mary Margaret* was owned by Bernard Taltavulli Biloxi Canning Company and would be piloted by Captain Willie Green. Captain Green stated: "I am into give the hardest fight to win the race. My boats plenty fast and should win."

The *Julia Delacruz* and the *Mary Foster* were owned by C.B. Foster Packing Company. The *Julia Delacruz* was considered a fast schooner and was a past champion of the regatta. Many locals felt the *Mary Foster* was the fastest of all the entries. Master boat builder Jack Covacevich built the *H.E. Gumbel* and the *I. Heidenheim*. He indicated that one of his two boats would win the race.

The *Daily Herald* of July 9 indicated that only four schooners would race for the Gulf of Mexico Title. The Biloxi harbor was the scene of the two-day water carnival, and all the boats had been colorfully decorated for the event. In addition to the race of the white-winged schooners, races were conducted for the catboat division. The catboats ranged from small sailing skiffs to the fish class fleet of the yacht club. In the larger sailing class division, competition would take place between the large sloops of the Southern Yacht Club. Finally, in the outboard motor division, races would be set up by class and one free-for-all race.

The *H.E. Gumbel* and the *I. Heidenheim* (both past winners) and newcomer the *Mary Margaret* were all built by master boat builder Jack Covacevich. The four captains were all veterans at schooner racing. Captain Green, captain of the *Mary Margaret*, had won with the *Jolly Traveler* in 1904 and the *Julia Delacruz* in 1924. Captain Martin Fountain, captain of the *H.E. Gumbel*, won with the *Henry M.* in 1910, 1912 and 1917 and with *Curtis Fountain* in 1920 and 1921. The *H.E. Gumbel* was the 1927 champion with Captain Fritz Santa Cruz. Captain Santa Cruz, a veteran skipper from North Biloxi, was at the helm of the *Julia Delacruz*. All were pitted against defending champions Captain Amos Ross and the schooner *I. Heidenheim*.

The *Julia Delacruz* was the only old clipper type in the 1924 race and had won in a gale. Many old-timers felt it would have the best chance in a good blow. The regatta was sponsored by the Biloxi Yacht Club and assisted by the Biloxi Chamber of Commerce. The *Daily Herald* indicated that many leading syndicated newspapers and magazine publications of the nation would be on two press boats.

On July 10, strong winds blew at the firing of the starting gun that officially kicked off the Thirtieth Gulf Fishermen's Schooner Championship Classic. Local and visiting yachts and crafts of all types lined the racecourse. Thousands of spectators dotted the beach, sea walls and piers. At the gun, the *H.E. Gumbel* took the lead, but the *Mary Margaret* cut inside on a starboard tack, forcing the *H.E. Gumbel* to back off. The *Mary Margaret* and *I. Heidenheim* continued neck and neck to the first turn beacon. As they continued down the channel, it became apparent that the *Mary Margaret* was stealing the wind from the *I. Heidenheim*, forcing it to back off to gain wind. For the first forty-five minutes, the race appeared to be evenly matched, with all three schooners making a good showing. The strong wind would become a deciding factor.

Halfway through the course the *H.E. Gumbel*'s port rigging on the main mast broke. This forced Captain Martin to withdraw the schooner from the race. The strong winds also had an effect of the *I. Heidenheim*, Captain Ross and his crew. On three separate occasions, precious minutes were lost due to

Schooners and a catboat sail down the Biloxi Channel during the annual regatta. *Courtesy of Alan Santa Cruz Collection.*

strong winds placing it behind the leader, the *Mary Margaret*. As for the *Mary Margaret*, the winds had little or no effect as Captain Green and his crew sailed a perfect race. After two hours and twenty minutes, the *Mary Margaret* crossed the finish line in first place. The *Mary Margaret*, Captain Green and his crew were presented the Holmhaven Permanent Trophy and $125. The second-place winner, *I. Heidenheim*, was awarded $50. The Gulf Fishermen's Schooner Championship Classic was completed, but the competition was not over; the action just shifted to the Pass Christian regatta. The *Daily Herald* of July 12 indicated that "fleets of sailing craft leave Biloxi today to compete in Pass Yacht Club Regatta."

Once again, the schooners *H.E. Gumbel*, *I. Heidenheim* and *Mary Margaret* were pitted against one another. They were joined by a flotilla of sailing and motoring vessels of all types from Biloxi. Once again, Captain Amos Ross was at the helm of the *I. Heidenheim* and Captain Martin Fountain was at the *H.E. Gumbel*. Due to illness, Captain Green was not at the helm of the *Mary Margaret*. Instead, Captain Bijou Tiblier, who won with the *H.E. Gumbel* in 1927, took Captain Green's place.

The race started with only a slight breeze, and the *I. Heidenheim* took the fifteen-second lead over the *Mary Margaret*. The *H.E. Gumbel* was forty-five seconds behind the *I. Heidenheim* in the eighteen-mile race. The *I. Heidenheim* continued to increase its lead while tacking toward the eastward stake. By the end of the first round, it was four minutes and fifteen seconds ahead of the *Mary Margaret*. During the second round, the wind increased to eighteen knots, and the *Mary Margaret* began to cut the *I. Heidenheim*'s lead. As they started the last round, the *Mary Margaret* was in striking distance. As the *Mary Margaret* attempted to pass the *I. Heidenheim*, both crews were laboring. The *Daily Herald* reported that in passing the *I. Heidenheim*'s bow on a port tack, the *Mary Margaret* lacked only about two feet on the extreme end of its main boom. At this point, the *I. Heidenheim* took advantage of its right of way on a starboard tack and hit the boom, only to suffer the consequences. The *I. Heidenheim*'s flying jib stay broke, forcing the *I. Heidenheim* to withdraw from the race. The *Mary Margaret* won the race after three hours and ten minutes.

A protest was registered by the *H.E. Gumbel* stating that due to the fact that both the *I. Heidenheim* and the *Mary Margaret* were involved in a foul, the *Gumbel* should be awarded the first place. The judges ruled that even though the *I. Heidenheim* might have been justified in hitting the *Mary Margaret* in accordance with marine rules for running in a race, no official protest was entered by the captain or a representative of that boat. The *Mary Margaret* was awarded first place and had won its second race of 1929.

Even before the 1929 race, factory owners and fishermen were converting old schooners from sail power to motor with sail backup. These schooners were called auxiliary schooners. The Biloxi Regattas of 1930 and 1931 featured races of these auxiliary schooners, but no sailing schooner raced. During the 1932 regatta, the sailing schooner division was once again added, and the *Mary Margaret* won again. In 1938, the last race of the white wings sailed past the Biloxi Yacht Club. The once graceful schooners that were not scraped were converted to carry gasoline and diesel engines with their rear masts removed.

For over fifty-five years, no race of the white-winged queens occurred, but that was about to change. In the 1990s, the Maritime & Seafood Industry Museum resurrected the schooner races. During the early 1990s, the museum built two replicas, the *Glen L. Swetman* and the *Mike Sekul*, of the old oyster schooners. The Biloxi Yacht Club continues to host sailing regattas. The Maritime & Seafood Industry Museum hosts its annual Gulf Coast Wooden Boat Show, where wooden yachts and schooners from this bygone era can be viewed.

Chapter 8

Blessing of the Fleet

The history of the Biloxi Blessing of the Fleet has deep ties to St. Michael Catholic Church. St. Michael has always been known as the "fishermen's church" due to the fact that everyone on the Point, in one way or another, was involved in the fishing industry. In addition to St. Michael, St. John in Biloxi and Sacred Heart in D'Iberville also had ties to the early blessing. The history of these three churches is rooted in the seafood industry and, more importantly, in the seafood families. These seafood families were the body of St. Michael, St. John and Sacred Heart Parishes. Their hard-earned dollars were given generously in support of these parishes.

During the 1880s, large numbers of immigrants arrived to work the seafood industry. Bishop Gunn, the bishop of the Catholic Diocese of Natchez, began to take notice of these seafood laborers and the fact that most were Catholic. He also realized that the spiritual and physical needs of the seafood families were not being cared for by the Mother Church. He urged Father Alphonse Ketels, pastor of Nativity BVM Church of Biloxi, to open a mission church and a school to be run and serviced by Nativity's priest. In 1907, St. Michael the Archangel Mission was opened on Point Cadet. This would be the beginning of the fishermen's church. In the fall of 1917, Bishop Gunn declared St. Michael a parish.

By the 1900s, Slavic, Bohemian and Cajun immigrants were living on the Point and Biloxi's Back Bay. In the early 1920s, Bishop Gunn turned his attention to Biloxi's Back Bay. Land was purchased from Mr. William Gorenflo. St. John Church was organized in 1921, and the new church was

The second St. Michael Catholic Church on Point Cadet was dedicated in August 1964 and is known throughout Biloxi as the "fishermen's church." *Courtesy of the Maritime & Seafood Industry Museum.*

dedicated on December 18, 1921. Father Patrick J. Carey was appointed pastor of St. John and Sacred Heart Parish. Sacred Heart Catholic Church parish had been established in 1859 in the Seymour community in North Biloxi, in the present-day city of D'Iberville and St. Martin communities. North Biloxi's fishing families were made up of descendants of colonial French and Spanish.

The arrival of these immigrants would also usher in many traditions that had their historical roots hundreds of years earlier in coastal European and Mediterranean communities. The communities of North Biloxi, Back Bay and the Point would come together to create one of the most beloved traditions, the Blessing of the Fleet.

The first blessings took place on Sunday, August 11, 1929. The *Daily Herald* reported that two "impressive ceremonies" were held on North Biloxi shore and Point Cadet. The D'Iberville blessing took place on the old Sacred Heart Church property just east of Central Avenue. The boats had been anchored on the outside of the island that sits south of the property. Father P.J. Carey, pastor of St. John and Sacred Heart Catholic Churches, conducted the ceremony. Hundreds had gathered on the shore to watch the

The 1929 twin blessing. In the top picture, Biloxi's Point Cadet blessing shows Archie Deslivey and Luke Dubaz to the extreme left; Sidney Forman, altar boy; Father McDonnell, missionary priest; Wilmont Boudreaux, altar boy; Father Hildebrand, pastor of St. Michael Church; Yankie Barhanovich and Edmond Boudreaux Sr., altar boys. The bottom picture shows Father Carey, his assistant Father Peter J. Quinn and Patrick M. Santa Cruz, one of the altar boys. *Courtesy of the Maritime & Seafood Industry Museum.*

various-sized vessels. Mass was celebrated on a specially constructed altar with both St. John's and Sacred Heart's choirs singing. It was noted that "Ave Maria" was sung by William J. Giblin and "Ave Maria Stella" was sung by Mr. Giblin and Dr. Seymour of the Veteran Hospital.

After the services, Father Carey and his assistant, Father P.J. Quinn, began blessing the vessels by walking from boat to boat. Unfortunately, Father Carey slipped during one of the crossings and dislocated his shoulder. He was carried to Biloxi Hospital, and Father Quinn completed the blessings.

On Point Cadet, the blessing ceremony was conducted by Father Hildebrand of St. Michael the Archangel Catholic Church. An altar had been set up on Dunbar Dukate factory pier. The St. Michael's choir was located on a boat tied to the pier. The church's organ had been moved to the pier and was played by Miss Marguerite Wink, accompanied by Judge J.B. Wink and George Dubaz on violins. Father Hildebrand celebrated the Mass with Father McDonnell, a missionary priest, assisting him. After the

A 1950s Blessing of the Fleet at Biloxi. *Back row*: Father Curley, Father Herbert Mullin, holding cross Foreman, Woodrow Guiher, unknown priest. *Front row*: Alan Santa Cruz, Armond Boudreaux, Kenny Foreman and Gray Toche. *Courtesy of the Maritime & Seafood Industry Museum.*

services, Father Hildebrand and Father McDonnell began going from boat to boat blessing all.

Through the years, the local celebration has changed in different ways, but the blessing of the boats, fishermen and their families has remained constant. The act of blessing the fleet has been accomplished in a variety of manners. During the earliest blessings, the priest actually climbed from boat to boat as they were assembled close to the shore or were rafted next to a pier. Today, the vessels pass in front of the blessing boat to receive the blessing.

Beginning with the first blessing in 1929, the local fishing fleet was blessed in both North Biloxi (present-day D'Iberville) and Point Cadet in Biloxi. Then, in 1947, a single event took place in the vicinity of the Coast Guard base on Biloxi's Point Cadet. Before being destroyed by Hurricane Katrina in 2005, the Point Cadet Plaza, the Maritime & Seafood Industry Museum and J.L. Scott Marine Education Center occupied this property. Yet the evolution was not completed until 1953. The blessing took its present-day form when a blessing boat was anchored adjacent to the channel off Biloxi's Central Beach and working boats paraded down the channel to be blessed.

Since the initial blessing in 1929, a number of festive activities have been added, some removed and others changed. These have included street parades, "faire do-do" dances, awards for best-decorated vessels and the naming of a king and queen. The translation of the faire do-do dance is from the words *faire* (to make) and *do-do* (to sleep). So it simply means to go to sleep. And what does that have to do with a street dance? We have to visit the Cajun culture in south Louisiana to get to the bottom of this mystery. The first "festival" took place in 1952. Then, in 1955, the first official Biloxi Shrimp Festival was held in conjunction with the Blessing of the Fleet. In south Louisiana during the horse and buggy days, private and public dances were held just about every weekend. Everyone from the grandparents to the small children attended the event. All brought food for a kind of potluck supper, which was followed by the dance. The elders would visit and gossip, but the babies were put to sleep, or "*faire do-do*." It was a chance for single young folks to socialize, but the young women and girls were strictly chaperoned.

Beginning in 1948, a Shrimp Queen was selected to reign over the Blessing of the Fleet and Shrimp Festival. The Shrimp Queen is chosen from contestants whose family is/was in the seafood industry and is sponsored by organizations that have strong ties to the seafood industry, local seafood processors and families. The first queen crowned during the 1948 festival was Gloria Meaut. Then in 1956, the first Shrimp King was chosen to rule with the Shrimp Queen. The king was chosen from men long associated with the seafood industry. The 1956 King and Queen were Luka Kuljis and Marjorie Dukate.

One of the Blessing of the Fleet's most solemn moments occurs during the casting of a wreath into the water in memory of deceased fishermen. In recent years, the wreath has been tossed into the Biloxi Channel from a hovering helicopter.

For as long as I can remember, the Blessing of the Fleet announced the opening of shrimp season in Mississippi waters. When I was a boy, this event occurred on the first of August and later moved to the first of June. I can remember serving as an altar boy for the event. Father Herbert Mullins, pastor of St. Michael Church, would conduct the blessing and was usually assisted by another priest. Every altar boy wanted to be picked for the blessing, but it was usually the older boys who were chosen.

Since many members of my family were fishermen, I have attended the event for numerous years. Most fishermen and their families would head to one of the barrier islands for a full day of fun. They would then return to join

The parade of shrimp boats during the annual Blessing of the Fleet. *Courtesy of the Maritime & Seafood Industry Museum.*

the parade and have their boats and families blessed. During the early years, the blessing was as important to the fisherman and his family as Christmas, New Year's, Easter and the Fourth of July. It started the new season with Almighty God's blessing and also was a time for family and friends.

The annual Blessing of the Fleet and Shrimp Festival has been sponsored by the combined efforts of the Parish of St. Michael the Archangel Catholic Church, the Slavic Benevolent Lodge, the Fleur de Lis Society and the City of Biloxi. To this day, the Blessing of the Fleet continues to be an important part of Biloxi's coastal culture and history.

Chapter 9
Hurricanes

Historically, the Mississippi Gulf Coast has been hit by hurricanes from the Native American periods to Katrina in 2005. Three of the worst storms to visit the Gulf Coast were the 1947 Storm, Camille in 1969 and Katrina in 2005. Hurricanes were not given a name until 1953. The year 1953 was the beginning of the practice of giving female names to all hurricanes. In 1979, male names were added to the list.

When examining a list of the most intense hurricanes to hit the United States between 1900 and 2005, we find that seven of them hit the Mississippi Gulf Coast. Many more grazed the Mississippi Gulf Coast as they made landfall in Louisiana and Alabama. The intensity of hurricanes is determined by categories, with one being the lowest and five being the highest. Those that hit the Mississippi Gulf Coast were Camille, 1969, a Category 5; the 1947 Storm, a Category 4; Katrina, 2005, a Category 3; Frederic, 1979, a Category 3; Betsy, 1965, a Category 3; and Elena, 1985, a Category 3.

At the time of the 1947 Storm, there was little or no knowledge of the course a storm may take. Most warnings would be issued as the storm was ready to come ashore. In 1969, when Camille hit, residents were warned about ten hours beforehand. In fact, it was predicted that Camille would hit Florida until it took a turn and aimed at the Mississippi Gulf Coast. One must remember that in 1969, storm predictions were really an educational guess and the technology was not as good as today's.

Those of us who have weathered such storms as the 1947 Storm, Camille or Katrina can paint a picture of pain, despair, loss, thanksgiving and determination. Everyone who has weathered any of these three storms can tell you a hurricane story.

THE 1947 STORM

By the beginning of the 1947 storm season, the Mississippi Gulf Coast had not been hit by a hurricane since 1916, a thirty-one-year lull. Due to this fact, many Mississippi Gulf Coast residents had not experienced a storm in their lifetimes, let alone a serious storm. On September 11, the *Herald* indicated that a severe tropical storm had shown up in the South Atlantic over 2,000 miles from Miami. On September 16, the *Herald* reported that the storm had hit the Bahamas with 160-mile-per-hour winds and appeared to be headed for Florida. On September 18, the *Herald*'s headlines read "Hurricane Rips Florida." The storm crossed Florida, entered the Gulf of Mexico and took aim on Mississippi and Louisiana. In the Louisiana marshes, messages were being dropped by small planes warning shrimpers of the approaching storm.

The 1947 Storm came ashore on September 19 as a Category 4 storm. A Category 4 storm has winds of 131 to 160 miles per hour and a tidal surge of thirteen to eighteen feet, and damage is extreme. About three o'clock in the morning, the winds were clocked at 50 miles per hour; by eight o'clock, they were near 100 miles per hour. The storm's eye passed over the Chandeleurs Islands and then over New Orleans. The Mississippi Gulf Coast was on the dangerous right-front quadrant and received severe damage from Pascagoula to Bay Waveland. During the height of the storm, all power and phone service was lost, leaving no means of communicating with the outside world. The storm surge had risen to thirteen feet, with waves several feet high during the height of the storm. The horrific winds and the power of the water were too much for framed structures. The Gulf Coast was isolated from rail traffic because 80 miles of the L&N Railroad was out of commission. The railroad bridges on Biloxi Bay and Bay St. Louis had been washed out. On Highway 90, both lanes were undermined in many places but especially the south lane. Numerous boats had washed ashore and lay within the debris of piers, businesses and homes.

Along Biloxi's beachfront, all wooden structures were flattened. Boats, pilings, piers, appliances, automobiles, furniture and more were scattered along the beachfront. In Biloxi's business district, only two structures on the south side of Highway 90 survived: the yacht club and the USO building.

On Point Cadet, the east end of Biloxi, the shrimp and oyster factories were reduced to piles of wood with shrimp trawlers broken against them. Factories damaged or lost on Biloxi's Point include Kuluz Brothers, Dubaz Brothers Shrimp & Crabmeat Company, Luka Kuljis Factory & Oil Wharf, Biloxi Seafood, Victory Packing Company, Moore Seafood Company and DeJeans. John Mavar Fishing & Packing Company suffered several hundred thousand dollars of damage. Taltavull's Biloxi Canning & Packing Company was under five feet of water. On Back Bay Biloxi, the factories of Houston Gollott, Weems Brothers, Mike Illich and Roy Rosalies were hard hit. Also on Biloxi's Back Bay, Covacevich Shipyard suffered heavy damage.

North Biloxi (present-day D'Iberville) reported damage along the northern shoreline to many homes, as well as numerous large trees being down. C.F. Gollott Seafood Company on the north shore suffered heavy damage. During the seventeen days of life of the 1947 Storm, the total damage was $110 million in 1947 dollars. Like many severe storms of our past, the 1947 Storm is now part of our Gulf Coast history. It would remain the center point of conversation for twenty-two years.

Hurricane Camille

On August 14, 1969, Hurricane Camille formed west of the Cayman Islands. By the next day, it was a Category 3 hurricane near western Cuba. On August 16, it was upgraded to a Category 5 hurricane on the Saffir-Simpson Hurricane Scale. It gained its Category 5 rating just before it hit the Mississippi Gulf Coast. Late in the evening of August 17, Camille smashed into the Mississippi Gulf Coast. The eye passed somewhere near Bay St. Louis and Pass Christian. Here the storm surge was about twenty-four feet. The eye of the storm and the east quadrant were the most dangerous side of the storm, and unfortunately, Biloxi was on that side. For about four hours, the Mississippi Gulf Coast was pounded by wind, rain and tidal surge.

One of the mixed blessings of Camille was that it was a fast-moving storm. Camille "dumped" over eleven inches of rain on the Gulf Coast during its landfall. The storm surge destroyed and flooded many businesses, historical structures and homes. On Biloxi Point and Back Bay, most of the seafood factories were destroyed. Many shrimp trawlers, pleasure craft and huge barges were lost or washed inland. In the Port of Gulfport, three large container ships were left high and dry on U.S. Highway 90. As the winds subsided and survivors ventured outside, the destruction was beyond their belief. Where once beautiful beachfront homes stood there were empty lots or the shells of damaged buildings. Huge piles of debris were the only evidence that once many homes and businesses existed here.

Electrical and phone service was destroyed. Many roads were impassable, and there was severe damage to several traffic and railroad bridges. Trees, power lines and vehicles were littered throughout the debris. Sections of Highway 90 appeared to have been combed with massive graters created by the turbulent surge of water.

Camille continued as a tropical depression through Mississippi, Tennessee, Kentucky, West Virginia and Virginia. Its winds over Biloxi had been clocked at over 210 miles per hour with a twenty-four-foot storm surge. In Harrison County, 68 square miles were destroyed. In 1969 dollars, the total damage was $1.42 billion—about $950 million of that along the Mississippi Gulf Coast. In Alabama, Mississippi and Louisiana, 153 were dead. The total death toll was 259, with 8,931 people injured. In addition, 5,662 homes were

In 1969, this tugboat was washed onto the sand beach by Hurricane Camille. *Courtesy of Maritime & Seafood Industry Museum.*

destroyed, and 13,915 had extensive damage. Add to this the loss of seafood factories, shrimp trawlers and pleasure craft.

All along our Gulf Coast, Camille became the benchmark for all storms. Those who weathered Camille felt we had seen the worst in our lifetime and nothing would or could ever compare to it.

Hurricane Katrina

On August 23, 2005, a new tropical depression formed near the Bahamas. Few people in the Biloxi Bay area noticed this twelfth storm of a very long hurricane season. On August 24, it became a tropical storm and was given the name Katrina. On August 25, just two hours before making landfall in Florida, Katrina became a Category 1 hurricane.

On August 28, Katrina was upgraded to a Category 5 storm with 175-mile-per-hour sustained winds and gusts of 215 miles per hour. It was now the fourth most intense storm in Atlantic storm recorded history. Katrina had caused thirty- to sixty-foot waves at sea, which, combined with the storm surge, would have devastating effects in the storm's northeastern quadrant.

Aftermath of Hurricane Camille shows the Episcopal Church of the Redeemer in Biloxi, 1969. *Courtesy of the Maritime & Seafood Industry Museum.*

During the early morning hours of August 29, Katrina weakened to a Category 3 storm and made its second landfall near the city of Buras, Louisiana. Within a few hours, Katrina would enter the Mississippi Sound and make its third landfall near the Louisiana and Mississippi state lines.

The Mississippi Gulf Coast was in the highly destructive northeastern quadrant. Katrina's intense eye wall winds and record surges began to smash and flood cities and communities. All over the Mississippi Gulf Coast, people began personal battles for their very survival.

Katrina had been an extremely monstrous storm moving so slowly that the people of the Biloxi Bay area were beginning to think it was never going to go away. Even though Katrina winds had not subsided, residents who were able began to venture out to sights that were unbelievable. A group of us had weathered the storm in BellSouth's old central office on Howard Avenue. In the early afternoon, we finally ventured out of our sanctuary. We walked east on Howard Avenue to the Vue Marche in downtown Biloxi. Then we headed south to Highway 90 and then west to the Biloxi Lighthouse. The farther we walked, the harder it became to comprehend what we saw. Our minds could not fathom the total destruction in every direction. In many places we climbed over piles and piles of debris only to discover that they were what was left of someone's home. Most of the beachfront homes in this section of Biloxi were totally destroyed. Most were nothing more than piles of splintered wood. As we headed west on Highway 90, we soon realized

In the aftermath of Hurricane Katrina, whole neighborhoods were destroyed throughout Biloxi. The destruction in this photograph shows a neighborhood just west of the Biloxi Lighthouse.

that the Biloxi Lighthouse was still standing—very battered but still standing. The Biloxi Lighthouse has long been the symbol of Biloxi and the Gulf Coast as well as its people, but now it became a symbol of hope.

Seeing this destruction showed us how powerful Mother Nature can be. What we saw made us realize that Hurricane Camille had truly become part of our past. Katrina's path of coastal destruction would extend from Louisiana to Florida but also reach deep into the interior of the Gulf Coast states. During the Gulf Wars, we heard a new military term: "shock and awe." The basic idea is to scare the enemy and create massive confusion. Among the Biloxi Bay communities' population, Katrina effectively created shock and awe, plus massive destruction.

On August 29, Katrina brought our world to a standstill. The citizens of the Biloxi Bay area all became part of living history. Yes, it would have a profound effect on our nation, Mississippi, the Gulf Coast and each individual for a very long time. The loss of human life was the hardest part of Katrina's rage. The death toll from Katrina would be about 1,836. In Mississippi it would be over 238, with 172 in the six coastal counties listed as dead or missing. Biloxi's death toll was 52 listed as dead or missing. Out of all the cities along the Mississippi Gulf Coast, Biloxi would have the highest loss of life. Yet one can only wonder how high the death toll could have been if Katrina had come ashore during the nighttime hours.

Along the Mississippi Gulf Coast, no one was left untouched. If their home or business was not destroyed, they suffered right along with their neighbors, business associates or family members who did lose everything. Hurricane Katrina was one visitor who came uninvited and stayed too long. Biloxi Bay area residents who lost their homes felt like they had gone through the death of a loved one. They went through all the emotions and then some. At first they felt numb and in disbelief, followed by hurting, anger, accepting the reality, depression and finally the determination to move forward with their lives.

Along Biloxi's Highway 90 and on the Point, every restaurant was destroyed or suffered severe damage. Historic restaurants like Mary Mahoney's Old French House, McElroy's Harbor House and the Ole Biloxi Schooner felt the wrath of Katrina. Mary Mahoney's Brunet-Fourchy House suffered severe damage. The Sidewalk Café, the Irish Pub and Court Yard buildings were destroyed. McElroy's Harbor House and the Ole Biloxi Schooner Restaurants were not just damaged—they were gone.

Katrina successfully removed childhood neighborhoods from the Biloxi Bay area. In D'Iberville and Biloxi, the oldest neighborhoods were destroyed. Katrina showed no prejudice as it destroyed the homes of the rich, middle class

The Biloxi Lighthouse and the 1999 Tri-Centennial statue of French explorer Pierre Lemoyne d'Iberville stand tall in the aftermath of Hurricane Katrina, 2005.

and poor. Katrina showed no mercy on churches, cultural centers, museums or businesses. The beautiful oaks and other trees, if not blown down, were stripped of all vegetation. All forms of wildlife felt the power of Katrina. Many were still stunned for hours and days after Katrina had passed.

The number of individuals left homeless was hard to humanly comprehend. Yet Katrina didn't stop there. Katrina tried to destroy our past by destroying historical structures that had become the fabric of our lives. Biloxi landmarks like Brielmaier House, Creole Cottage, Church of the Redeemer, Dantzler House, Father Ryan House, Fisherman's Cottage, Foretich House, Green Oaks Home, Old Santini House, Pleasant Reed House, Tullis Toledano Manor, Wesley House and many more were reduced to piles of wood and brick. Out of 160 landmark structures, 60 were lost in Biloxi. Beauvoir, Biloxi Lighthouse, Creole Cottage, Old Biloxi Cemetery, Old Biloxi Library, Old Brick House, Magnolia Hotel and St. Michael Church suffered unbelievable damage. The Maritime & Seafood Industry Museum was destroyed. The Ohr O'Keefe Museum suffered heavy damage, and the list goes on and on. Amazingly, the historic Ring in Oak on the Church of the Redeemer

In the aftermath of Hurricane Katrina, St. Michael (the fishermen's church) was surrounded by debris.

property survived. One could say Katrina destroyed our history, but our history is more than buildings—it is what gives us a sense of place.

As Katrina was approaching Biloxi, many shrimp boats and pleasure craft were moved to "safe harbor" in Biloxi Bay and up its river. Most would be destroyed, sunk or washed a long way from the water. The casino barges on Casino Row (old cannery row) were destroyed and washed ashore. All of the casino hotels in Biloxi were severely damaged. The casino barges on Biloxi Back Bay also suffered severe damage. The seafood factories in Biloxi Bay were destroyed or severely damaged. From the time Katrina formed on August 23 until it dissipated on August 30, its path of destruction in 2005 dollars cost over $81 billion.

For centuries, Biloxians have enjoyed their lives along the Mississippi Gulf Coast, living, working and playing in the waters of Biloxi Bay and the Mississippi Sound and enjoying the bountiful seafood harvested from Gulf waters. As with past storms, Biloxi and Gulf Coast natives have dug out of the storm debris to rebuild and create a future for their generation and generations to come.

Chapter 10

Biloxi's Historic Restaurants

Ask anyone who grew up along the Mississippi Gulf Coast and they will tell you it doesn't take long to figure out that our area offers a wide variety of food. The abundance of seafood due to our location is one factor. But in addition to seafood we have a wide variety of meats, vegetables and fruits. A long time ago, we could not pick the day our child would be born and as small children you could not pick the food you would eat. We were always served something to eat, and if you did not eat it, you went hungry (for one night anyway). As we grew older, we realized we could either eat to live or live to eat. Basically, most of us would say we live to eat. We enjoy eating what we consider good old southern Mississippi-Louisiana food with a Gulf Coast twist. Most of us grew up on good home-cooked meals and still have fond memories. We also grew up with certain locally owned restaurants that, if your family was fortunate, you visited frequently. Some of these restaurants have become part of our past and Gulf Coast history.

During the 1960s, many of us frequented places like Gus Stevens, Baricev's, Mary's Drive In, Sinopoli's, Patio Inn, Hugo's, Gumbo House, Alamo Fried Chicken, French's Restaurant, Speedway Café and Tony's Spaghetti House. For our fast-food needs, we would visit drive-ins like Big R Root Beer Drive Inn, Dunnaway's Drive Inn, Frostop, Lane Drive In, Teens Drive In or McElroy's A&W Root Beer, or small neighborhood cafés like Rosetti's Café, Pitalo's Café, East Inn Grill, Abbie's Little Diner and Marguerite's Café.

Most of these have disappeared into our past and only remain as memories, while a few have evolved from their former selves into something else over time. So let's take a look at four of these historic restaurants that became part of the fabric that made the Mississippi Gulf Coast a very special place to live and eat. Let's look at the evolution from Abbie's Little Diner to Burger-Burger; Mary Mahoney's Old French House Restaurant; McElroy's A&W Root Beer to McElroy's Harbor House Seafood Restaurant; and Rosetti's Café to the Ole Biloxi Schooner Restaurant.

Let us begin with Abbie's Little Diner. Abbie and Mildred Brunies of New Orleans moved to the city of Biloxi in 1945. Sometime between 1945 and 1949, they opened a small diner on the corner of LaSalle Street and Benachi Avenue in Biloxi. The diner was typical of the period, with a long bar and a few small tables with the grill in plain view of the patrons. The difference from other diners was that Abbie's had a po'boy called the burger-burger, which was the personal creation of the Brunies family. The burger-burger po'boy was filled with homemade hamburger patties, gravy, mustard, raw onions and the Brunies special sauce. The burger-burger became a local favorite, as well as Abbie's Little Diner's trademark po'boy. If you could eat a whole loaf of French bread you could eat a burger-burger, because that is how big it was. There was nothing like it anywhere around.

Albert "Abbie" Brunies was born into a musical family in 1900 in New Orleans. The Brunies family had emigrated from Switzerland to New Orleans. He and his six siblings learned to play musical instruments at young ages. Abbie played the trumpet and drums. Two of the groups he played with were the Four New Orleans Jazz Babies and New Orleans Rhythm Kings. Abbie's bands performed at the Halfway House in New Orleans from 1919 to 1927. In 1925, his music was recorded by Okeh Records and Columbia Records. Columbia Records would make additional recordings of Abbie's band in 1926, 1927 and 1928. He continued to play in the New Orleans area until 1945, when he moved to Biloxi. In Biloxi, he formed a new band and began playing at Baricev's Restaurant. He would also play his music with Johnny Bertucci and Bertucci's Melody Boys. In 1957 Biloxi, the Brunies Brothers Dixieland Jazz Band was recorded on the American Music label for the last time.

Even though the burger-burger was a local tradition, I was not introduced to it until after I was discharged from the navy and returned to the local workforce. Once I tasted my first, it definitely was not my last. The Brunieses were a husband-and-wife team, with Abbie being the more social one while Mildred was the more business-minded one. Of course, this combination turned into little spats in front of patrons or, as we perceived it, our personal reality show. These spats never really got out of hand, but one could look at it as dinner with free entertainment.

Burger-Burger, Howard Avenue, Biloxi.

The Brunieses' Abbie's Little Diner continued its operation on the corner of LaSalle Street and Benachi Avenue. In 1987, Martha Ann Ebberman purchased Abbie's Little Diner and the burger-burger trademark recipe. She continued operating at that location until 1993. Even with the closing of the diner, she continued serving the burger-burger in an Ocean Springs restaurant. During this period, she looked and hoped for a location in Biloxi for the return of the burger-burger.

In late 2000, Martha Ann and Jamie Sablich began conversing about a business venture to return the burger-burger to Biloxi. In January 2001, it became reality when Burger-Burger's doors opened at 1039 Howard Avenue in Biloxi. In no time at all, the burger-burger po'boy became a local favorite again, and the restaurant became a tourist stop. The future looked good until Hurricane Katrina came ashore on August 29, 2005. Though the building was not severely damaged, Burger-Burger, like so many businesses, began a struggle to open its doors and survive Katrina's aftermath. In mid-September, Burger-Burger became the first local restaurant to reopen its doors on the Biloxi Peninsula. Many of us working in Biloxi's downtown area would patiently wait our turn to enjoy the savory dishes as well as the burger-burger po'boy.

Most locals and tourists who have frequented Mary Mahoney's Old French House Restaurant would tell you it has beauty and allure with a touch of class, a touch of eloquence and a touch of Ole New Orleans with an Irish twist. All of this came from a dream, which then became a vision and eventually a reality of Mary Mahoney, a Biloxian of Yugoslavian descent.

Before Hurricane Katrina, like many locals, one of my favorite coffee spots and eateries was in the Old French House and Le Café. While I am on a roll, I must also mention there was nothing like having drinks in the Irish Pub, a favorite spot for many during Mardi Gras, St. Patrick's Day and other special local and personal events. In the Old French House, I have dined for anniversaries, birthdays, special occasions and business and social events. When you wanted to celebrate, it was just a special place.

Le Café will always be a special place to me. I have eaten many New Orleans–style po'boys there and had enough cups of coffee to last a lifetime. The coffee drinking was special due to the fact that so many locals would gang up in the café. Local businessmen, managers and employees, from blue-collar to white-collar workers, would all share their point of view on work, sports events, local events and national events. The conversations were sometimes borderline arguments, but they always remained civil. If only they could have had the ears of the politicians, they could have solved all the local and world problems.

The center point of Mary Mahoney's is the Brunet-Fourchy House. During the 1800s, many New Orleans families purchased property along the Mississippi Gulf Coast. Francois Gustave Brunet was one of them. He purchased the future site of Mary Mahoney's in 1835 for $800. Brunet retained the property for one year and then sold it to Alexander Francois Fourchy, a New Orleans attorney, for $1,800. The deed indicates that a house with four apartments and a front and back gallery was the reason for the higher sales price. The Fourchy family retained the property until 1850, when they sold it to Daniel Goos, who within a year sold it to Samuel Friedlander from New Orleans.

The property would continue to change hands, but in 1893 it was converted into a boardinghouse with additional rooms in three adjacent buildings. Between 1909 and 1962, it was once again a private residence. The Brunet-Fourchy House, though reoriented, is typical of the Creole cottage of New Orleans. The house is a four-room structure with two chimneys. Each chimney has a double fireplace, with one fireplace facing each room. The gallery has iron railing similar to other New Orleans structures. One of the interesting aspects of the house is the brickwork. The bricks were made by William Kendall Brick Works, which was in operation during the 1840s and early 1850s. Kendall's brickyard was located west of I-110 along the north shore of present-day D'Iberville. Two other Biloxi houses that were built with Kendall bricks were the Tullis Toledano Manor and the Old Brick House.

Mary Mahoney's Restaurant, Rue Magnolia, Biloxi.

To tell the story, we must look at the lives of the three individuals who would create Mary Mahoney's. Mary, her brother Andrew and their sister Anne lived at 1219 Howard Avenue in east Biloxi. Their parents were Anthony "Tony" and Mary Cvitanovich. Tony was the owner of Sanitary Fish & Oyster Company in Biloxi. Robert "Bob" Francis Mahoney, a Pennsylvania native, arrived in Biloxi courtesy of the U.S. Army Air Corps during World War II. He was stationed at Keesler Field, which would later become Keesler Air Force Base with the creation of the United States Air Force on September 18, 1947. After Bob Mahoney severed his ties with the service, he remained in Biloxi, where he met and married Mary Cvitanovich.

Bob Mahoney would be the first to become involved in Biloxi's tourism. Bob became a manager at the Broadwater and Tivoli Hotels. It would be Bob who would introduce Mary Mahoney to her true calling and the beginning of a legacy. Bob was asked by Tivoli management if Mary could manage the Tivoli Lounge, which she did, with some hesitation. Between 1956 and 1965, Bob worked for the *Biloxi-Gulfport Daily Herald*.

In 1962, the property and Brunet-Fourchy House were purchased by Mary Mahoney, her husband, Bob, and her brother, Andrew Cvitanovich. Mary Mahoney's dream of creating a fine dining restaurant was almost a reality. The doors of Mary Mahoney's Old French House Restaurant opened on

Ascension Thursday, May 7, 1964. Within three years, the restaurant had been featured in *Food Times* magazine, and soon more feature stories followed, and thus the legacy had begun. Mary Mahoney would become the first woman to receive the Small Business Person of the Year award in 1985. Another first was her election as the first female president of the Biloxi Chamber of Commerce.

Tragedy struck the Mahoney-Cvitanovich families too soon when Mary passed away on December 29, 1985. Yet the legacy would continue to grow. Bob Mahoney and Andrew Cvitanovich continued Mary's traditions with the assistance of family and friends. By 2005, like so many other Biloxi businesses, the future of Mary Mahoney's Old French House looked good. But a double tragedy struck the Mahoney-Cvitanovich families and their beloved Mississippi Gulf Coast. Bob Mahoney died on August 25, 2005, and just four days later, on August 29, Hurricane Katrina devastated the Mississippi Gulf Coast. The Mahoney-Cvitanovich families, in deep mourning, would not be spared this devastation. The Old French House suffered severe damage. The Sidewalk Café and the Irish Pub were destroyed. The Mahoney-Cvitanovich families began the tedious task of rebuilding and restoration. On November 3, 2005, the doors of Mary Mahoney's Old French House were open again.

Today, Mary Mahoney's legacy is carried on by Mary's children, Bob Jr. and Eileen; Mary's grandchildren, Stacy, Caty, Trae, Nicole and Josh; and Andrew Cvitanovich and his children, Andrea, JoAnna and Tony.

If you lived on the Gulf Coast before Hurricane Katrina, you most likely visited McElroy's Harbor House Restaurant. Like other local restaurants, McElroy's not only attracted locals but was a favorite of tourists as well. As at Mary Mahoney's, I have had quite a few cups of coffee in McElroy's Harbor House, as well as many seafood dishes. Also like Mary Mahoney's, the coffee drinking was special due to the fact that so many locals would gang up during the early morning hours. Employed and retired individuals would armchair quarterback and discuss the ways to solve the problems of the world, if just someone would listen. These armchair coaches and armchair politicians dissected everything from high school football to the New Orleans Saints. From local to national events, they voiced their opinions, which felt good but in the long run changed nothing.

The McElroys' involvement in the restaurant business goes back to the 1950s and the A&W Drive-In on Gill Avenue in Biloxi. Most Biloxians, at least the baby boomers, can remember the drive-in restaurants that dotted the Mississippi Gulf Coast. These drive-ins became the teen social hubs in most communities. Of course, many hamburgers, French fries, root beer floats and cherry cokes were consumed along the way. I can remember many visits to the

McElroy's Harbor House was destroyed by Hurricane Katrina, but plans are underway for it to be rebuilt in the near future. *Courtesy of Alan Santa Cruz Collection.*

A&W Drive-In to socialize with friends and family. I can still taste the flavor of the A&W root beer. The glass mugs were so cold the root beer would almost freeze to slush before you could start drinking it. There was nothing like it then, and there is still nothing like it. They even had baby-size mugs for kids.

So let's look at the McElroy family and a business that started as a second income. James Ransom Sr. and Irene McElroy moved to Biloxi from Columbus, Mississippi, sometime around 1913. In 1913, James Sr. began working at Biloxi Machine and Equipment Company, which was originally owned by Dr. Albert Babendreer. The 1927 city directories show James McElroy and William "Bill" Vierling as being employed by Dr. Babendreer. In the mid-1930s, James McElroy, J.H. Osborne and William "Bill" Vierling purchased the company and changed its name to Biloxi Machine Works. In 1949, James McElroy Sr. took ownership, and the name was changed to McElroy's Machine Works.

James and Irene lived on Gill Avenue in Biloxi and continued operating McElroy's Machine Works until about 1967. Their son James Jr. was born in Biloxi and married Emerick Redman from Gulfport. James Jr. went to work for the U.S. Postal Service, where he would work for thirty-six years before retiring. It was James Jr. who took the McElroy family in a different direction—the restaurant business.

In 1954, on the corner of Gill Avenue and West Howard Avenue (today's Judge Sekul Drive), was Trevor's Drive In. R.M. and Cora Trevor were listed as the owners, but Cora was listed as a widow in 1954. Around 1958, James and Emerick McElroy purchased Trevor's Drive In and converted it to an A&W Drive-In. Even though James continued working for the postal service, the drive-in became a second source of income to the McElroy family.

In 1973, the McElroy family started a second A&W Drive-In at 461 Central Avenue in North Biloxi (present-day D'Iberville). They also took over the Biloxi Small Craft Harbor's snack bar. The Biloxi Marina Snack Bar was under the management of Mickey McElroy, son of James and Emerick. This would bring about the creation of James R. McElroy Inc., owned by James Jr. and Emerick McElroy, Mickey McElroy and his wife, Jane. By 1975, the Biloxi Marina Snack Bar was evolving into something more than just a snack bar. The McElroy family decided to close the two A&W Drive-Ins in 1975.

Then, in February 1976, McElroy's Harbor House restaurant opened its doors. During the next thirty years, McElroy's Harbor House Restaurant became a favorite eating establishment of locals and tourist alike. In 2003, the McElroy family expanded their business with the opening of McElroy's on the Bayou in Ocean Springs. Sadly, the family would lose both James Jr. and Emerick before 2005. In 2005, the future looked good for Mickey and Jane McElroy and our Mississippi Gulf Coast. Then, on August 29, the unthinkable occurred when Hurricane Katrina slammed into the Mississippi Gulf Coast. McElroy's Harbor House was not just damaged—it was removed as if it never existed. McElroy's on the Bayou had seven feet of water that damaged the building and destroyed all its furnishings.

McElroy's on the Bayou would take six months to remodel and replace all its furnishings, but McElroy's Harbor House is a different story. In 2011, the McElroy family are planning and working on the rebuilding of McElroy's Harbor House. Hopefully, in the near future we will be able to once again enjoy the view of sailboats and shrimp boats on the Mississippi Sound while dining on seafood in McElroy's Harbor House Restaurant.

Growing up on the Point, one of my fondest memories was going to Rosetti's Café to eat a po'boy called a Vancleave Special and wash it down with an ice cold Barq's Root Beer. I know that I am not the only one who has similar fond memories of Rosetti's Café. Even though Rosetti's Café and the Biloxi Schooner were two separate businesses, most locals think of them as one and the same. So let's look at the history of these two businesses and what common ground they share.

In the 1930s, the corner of Myrtle Street and Howard Avenue had turned into thriving businesses. The family of Jake and Pauline Rosetti were operating Rosetti's Service Station and Rosetti's Beer Garden. Between 1936 and 1941, the family expanded their business by adding Hi-Way Grocery and

Rosetti's Tourist Court. Rosetti's Beer Garden was most likely established in about 1933 and would later become Rosetti's Café. The 1949 City of Biloxi Directory lists Rosetti's Café as owned by Vincent J. and George Rosetti, two of Jake and Pauline's sons. By 1952, George had joined the Biloxi Police Department, and only Vincent was listed as the owner. Vici, as Vincent was known by locals, would be the first to introduce New Orleans–style po'boys with his own personal touch to the Mississippi Gulf Coast.

A po'boy is a sandwich made with French bread that can be filled with fried oysters, shrimp, fish, soft-shell crabs, crawfish, roast beet with gravy, roast pork, meatballs, smoked sausage and on and on. Clovis and Benjamin Martin, who were former streetcar drivers in New Orleans, were credited with inventing the po'boy. The Martin brothers operated a coffee stand and restaurant on St. Claude Avenue in the French Market. During 1929, the transit strike in New Orleans turned violent, with lives lost, trolleys burned, buildings destroyed and other major destruction. The Martin brothers took pity on these poor boys and former colleagues from Division 194. They offered a free sandwich filled with gravy and roast beef or gravy and sliced potatoes. It was said that when a striker would come by, the brothers would announce "another poor boy," thus the birth of a po'boy.

It has been reported that Vici knew the Martin brothers, and he would offer a similar menu but personalized. Vici would hand press the po'boys on the grill, creating a mouth-watering hot sandwich. In no time, Rosetti's became a favorite of locals and tourists alike. Many locals would travel from many locations along the Mississippi Gulf Coast just for a taste of Vici's po'boys. One cold November night in 1946, four young men from Vancleave made the journey to their favorite eating place. Norris Jones, Jason Mallette and Roy B. Roberts placed their order with Vici. Vici turned to C.M. "Kipp" Dees, whom he called Vancleave, and asked what he wanted. After telling Vici that it wasn't on the menu, Vici indicated that they would make it, whatever it was. Kipp then ordered a crabmeat po'boy with cheese. Vici added the new po'boy to Rosetti's menu and called it a Vancleave Special.

On September 19, 1947, a Category 4 storm struck Louisiana and Mississippi's Gulf Coast. Like many businesses and homes, the original Rosetti's Café suffered heavy damage. Yet like so many Biloxians, Vici rebuilt Rosetti's Café and continued to run it until his health began to deteriorate. In 1969, he sold Rosetti's Café and his secret recipes, including the Vancleave Special, to John Lewis, a longtime employee of Vici who began working at Rosetti's Café when he was thirteen years old. John Lewis purchased Rosetti's Café in July 1969. Then, on August 17, 1969, Hurricane Camille struck the Mississippi

Ole Biloxi Schooner Restaurant, Howard Avenue, Biloxi.

Gulf Coast. Once again, Rosetti's Café suffered severe damage. John Lewis rebuilt and, like Vici, continued the longtime Biloxi tradition of Rosetti's Café.

In 1983, John Lewis built a new building on Caillavet Street in downtown Biloxi and relocated Rosetti's Café. After relocating Rosetti's Café, he sold the property and building on the Point to Joe Lancon. John Lewis continued Rosetti's Café at the Caillavet Street location until 1990. Like Vici before him, for health reasons, the doors of Rosetti's Café closed for the last time and faded into history.

In December 1983, a new tradition was born when the doors of the Biloxi Schooner Restaurant opened. Joe Lancon had taken over the former location of Rosetti's Café. Like Rosetti's Café before him, Joe offered his own version of New Orleans po'boys. Like Rosetti's, the po'boys were pressed on the grill.

After the closing of Rosetti's Café, its patrons began to visit the Biloxi Schooner. Joe Lancon and his wife, Charmaine, continued the Biloxi Schooner tradition on the Point until August 29, 2005. Hurricane Katrina, like the 1947 Storm and Camille in 1969, destroyed the building and devastated east Biloxi. Joe Lancon made a decision to relocate the Biloxi Schooner to downtown Biloxi. In June 2006, the doors of the Biloxi Schooner reopened in a new location. Though larger than the old location, the Lancon family remodeled the building to resemble the old one. Once again, loyal as well as new patrons and tourists returned to sample this old Biloxi tradition.

Chapter 11

End of an Era

After Hurricane Camille, the economy of Biloxi recovered, but slowly. The return to normal after Camille included the return of many events. Many Gulf Coast cities sponsored Mardi Gras parades. In March, the annual Spring Pilgrimages were sponsored by local garden clubs. Biloxi's annual Blessing of the Shrimp Fleet was held the first weekend in June. During the summer months, local yacht clubs held sailing regattas, and fishing rodeos became very popular. The business and tourism communities began the long recovery. The return to normal after Camille was slow but the improvement apparent.

In 1976, the Vietnamese began arriving, and in no time their community began to flourish. Most would become laborers or fishermen in the seafood industry. Like immigrant families before them, the whole family was involved in the seafood industry. From 1970 to 1990, the seafood industry witnessed various new technologies and developments to improve the processing of seafood. The evolution from canning to flash freezing was complete. New shrimp-picking machines and other innovations reduced the number of factory workers. Then, in 1992, the winds of change began to affect the city of Biloxi.

In 1992, dockside gambling was legalized in Mississippi. A new industry had arrived in Biloxi that would change the face of Biloxi's Point and Back Bay. The Isle of Capri Casino was the first to open its doors on old cannery row. Soon, another section of cannery row was acquired by Lady Luck Casino. Cannery row was becoming Casino Row. In a short time, cannery

Biloxi shrimp boats line the piers on Back Bay Biloxi, March 2011.

row had become the home of Isle of Capri Casino, Casino Magic, Grand Casino and Lady Luck Casino. The seafood industry was now located on Biloxi Bay's north shore and Back Bay. After 1999, Biloxi appeared to be on a fast growth path with nothing to slow it down. In the late 1990s, Biloxi was the site of nine casinos: the Beau Rivage, Boomtown, Casino Magic, Isle of Capri, Grand Casino, Imperial Palace, New Palace, President and Treasure Bay.

The success of gaming in Biloxi had a wondrous effect on all elements of Biloxi. The addition of these casinos helped push Biloxi Bay communities and the Mississippi Gulf Coast into the next chapter of the gaming and tourism industries. It wasn't just the gaming that was attracting tourists; the rich local history and culture of the Biloxi Bay communities were helping to promote the Gulf Coast as one of the top tourist destinations. With the Gulf Shore's white sand beaches, recreational boating, fishing, deep-sea fishing, sailing, windsurfing, jet skiing, boat races and other water activities, Biloxi had so much to offer. The Biloxi Bay area has historically been the home to various golf courses, and in 2000 there were twenty courses in Biloxi and surrounding areas. Add to the mix a little lagniappe in the form of fresh seafood from the Mississippi Sound and other local cuisine, and everyone would agree that life was good.

Built in the 1850s, the Old Brick House can be seen after Hurricane Katrina in 2005 and after restoration in 2011.

The 2000 census indicated that Biloxi's population was 50,644, ranking it as the third largest in Mississippi. With new growth and the success of gaming, the Mississippi Gulf Coast began to see more winds of change. In Biloxi Bay communities, all of this led to new jobs in tourism, hospitality, gaming, gaming support businesses, new construction, related trades and many other technical and labor-related fields. In 2000, Biloxi's unemployment rate was 4 percent, compared to national figures of 5.5 percent and Mississippi's 6.4 percent. The hotel and motel industry also began to benefit from Biloxi's nine casinos. The no-vacancy signs became the normal on any given weekend.

The 2000 census also indicated that individuals in the Biloxi Bay communities were working in a wide variety of management and professional fields. These industry fields were in agriculture, forestry, fishing, hunting, construction, manufacturing, wholesale trade, retail trade, transportation, warehouse, utilities, information, finance, insurance, real estate, rental, leasing, professional, scientific, administrative, waste management services, educational, health, social services, arts, entertainment, recreation, accommodation, food services, public administration and other related fields. The shrimpers and seafood factory workers were still making a good living in Biloxi's historical seafood industry. The seafood industries' products were still a tourism and local draw.

The Biloxi Bay communities were also feeling the effects of the success of gaming in Biloxi. Public works projects were in full swing, with expansion of transportation infrastructures, recreational and educational facilities all getting much-needed attention. Harrison County was also busy with various road projects and recreational and educational facilities. Jobs were plentiful in many and various fields.

In D'Iberville, development began to take place in an area that would become Sangani Boulevard. In time, various restaurants and shops began to appear. The Biloxi beachfront was dotted with more than sixty various restaurants. Some of the restaurants were incorporated into the casino resorts, some in hotel or motel complexes and some in Edgewater Mall and

St. Michael after Katrina in 2005 and after restoration in 2011.

Edgewater Village shopping centers. There was everything from fine dining to fast food, with some being chain restaurants while others were privately owned. On any given weekend at any restaurant, it was not uncommon to wait a few hours to be seated. Biloxi's Gulf waters and its selections of seafood made it a very special place to play, live and eat.

In Biloxi's historic districts, many historical homes and businesses were on the National Register of Historic Places. In fact, along the Mississippi Gulf Coast, historical preservation was on the minds of each community. In Biloxi, casino tax revenues were being used for historic preservation. Places that benefited from such revenues were Beauvoir, Biloxi City Hall, Biloxi Lighthouse, Brielmaier House, Creole Cottage, Church of the Redeemer with the historic Ring in Oak, Dantzler House, Father Ryan House, Fisherman's Cottage, Foretich House, Gillis House, Green Oaks Home, Magnolia Hotel, Old Brick House, Old Biloxi Cemetery, Old Biloxi Library, Mary Mahoney's Old French House, Old Spanish House, Old Santini House, Pleasant Reed House, Redding House, St. Michael's (the fishermen's church), Tullis Toledano Manor and Wesley House.

Biloxi truly was a historic city, with historic structures in every section. On the Point there were 22 historic structures; in the east central district there were 42; in the downtown district there were 30; in the central district there were 42; in Back Bay there were 10; in the west central district there were 40; in the West Beach district there were 29; in West Biloxi there were 19; and in the North Biloxi and Woolmarket area there were 6. Of these historical structures, 160 were landmarks.

Biloxi's cultural points of interest included the French Colonial Burial Grounds, located at Moran Art Studio on Porter Avenue. Biloxi was home to some upcoming and overall great museums. The Ohr O'Keefe Museum was in the process of creating a world-renowned art museum and community. The Maritime & Seafood Industry Museum had developed into a first-class historical and educational museum. The Mardi Gras Museum had just moved into the Historic Dantzler House near Biloxi's Historic Lighthouse. There was also the J.L. Scott Marine Education Center and Aquarium on Beach Boulevard east of Casino Row. In August 2005, with new growth and new construction in tourism, local industry and revenue from a bustling gaming industry, the future of the Biloxi area was good. But the fair winds would begin to change.

On August 29, 2005, Hurricane Katrina struck the Mississippi Gulf Coast, and we all became part of history. Even though it is our recent history, it will have a profound effect on our nation, Mississippi, the Gulf Coast and each

The Hurricane Katrina Memorial on the Biloxi Town Green depicts Hurricane Katrina's eternal waves and the lists of the Mississippi Gulf Coast victims and missing.

individual for a very long time. No one in Biloxi or the Mississippi Gulf Coast was left untouched.

Along the Mississippi Gulf Coast, communities began to remove debris and rebuild after the storm. In Biloxi, recovery was a struggle to rebuild the infrastructure. Basic utilities had to be restored where it was possible and some areas abandoned until a future time. By early 2007, the seafood processing plants and shrimpers were struggling but making progress. In the casino industry, Beau Rivage Casino, Boomtown Casino, Grand Casino, Hard Rock Casino, IP Casino Resort & Spa, Palace Casino Resort and Treasure Bay Casino Resort were opened for business, and tourists were beginning to return. The seafood and tourist industries were beginning to make progress. Fishing and sailing once again became weekend pastimes. New construction and restoration of homes, businesses and government structures were visible everywhere.

In 2008, the Wall Street crisis began and festered. It would be a long, hard road to recovery. In Biloxi and along the Mississippi Gulf Coast, the hurricane recovery money was aiding the local economy. As the Wall Street crisis continued, the local economy began to wane. In Biloxi and the Mississippi

The 163-year-old Biloxi Lighthouse after restoration and repairs from Hurricane Katrina damage, March 2011.

Gulf Coast, tourism began to slow down. The next two years would be a struggle of balancing budgets and just surviving tough times.

In early 2010, it appeared that tourism was improving. Then the unthinkable occurred. On April 20, 2010, the Deepwater Horizon oil-drilling rig exploded, causing millions of gallons of oil to spill into the Gulf. To the people of Biloxi and the Mississippi Gulf Coast, this was a third crisis. Still recovering from Hurricane Katrina and the economic crisis, they now faced this new nemesis. In late 2010, the recovery from the Economic Crisis and Gulf Oil Spill began to show signs of hope. The seafood processors, shrimpers, restaurants, casinos and tourism struggled with and tried to recover from this third disaster.

In 2011, the scars of Katrina are still visible in Biloxi and along the Mississippi Gulf Coast. The signs of new construction and the completion of new buildings continue to add to Biloxi's revival. Even though recovery appears to be slow, it is very steady. Like a child taking small steps, our steps are becoming larger as we rebuild together. The people of Biloxi and the Mississippi Coast, like past generations, are making a difference one day at a time. It matters not if you are a new resident or one whose heritage has tied you to Biloxi's maritime and seafood industry for generations. In 1912, it was noted by a visitor that Biloxi had two principal crops, "shrimp and people," a statement that is very true to this day. Today, Biloxi's revival is tied to its people, who in turn are tied to its maritime and seafood industry history.

References

Books and Articles

Bergeron, Kat. *Marine Resources and History of the Mississippi Gulf Coast*. Vol. 1, *History of Coast Lighthouse*. Jackson: Mississippi Department of Marine Resources, 1998.

Blitz, John M., and C. Baxter Mann Jr. *Fisherfolk, Framers, and Frenchmen: Archaeological Explorations on the Mississippi Gulf Coast*. Jackson: Mississippi Department of Archives and History, 2000.

Brasseux, Carl A. *A Comparative View of French Louisiana, 1699 and 1762*. Lafayette: University of Southwestern Louisiana, 1979.

Brasseaux, Carl A., and Glenn R. Conrad. *The Last Years of French Louisiana by Marc de Villiers du Terrage*. Trans. by Hosea Phillips. Lafayette: University of Southwestern Louisiana, 1982.

Burt, Jesse, and Robert B. Ferguson. *Indians of Southeast: Then and Now*. Nashville, TN: Abingdon Press, 1973.

City of Biloxi. *The Buildings of Biloxi: An Architectural Survey*. Biloxi: City of Biloxi, Mississippi, 2000.

Coker, William S., ed., Jerome F. Coling, cartographer, and Index by Polly Coker. *The Military Presence on the Gulf Coast*. Pensacola, FL: Gulf Coast History and Humanities Conference, 1978.

Conrad, Glenn R. *Immigration and War Louisiana 1718–1721, from the Memoir of Charles Le Gac translated, edited and annotated*. Lafayette: University of Southwestern Louisiana, 1970.

Deiler, J. Hanno. *The Settlement of the German Coast of Louisiana*. Philadelphia: American Germanic Press, 1909.

De le Harpe, Jean-Baptiste Benard. *The Historical Journal of the Establishment of the French in Louisiana*. Lafayette: University of Southwestern Louisiana, 1971.

Foreman, Grant. *The Five Civilized Tribes*. Muskogee, OK: Press of Hoffman Printing, 1966.

French, Benjamin Franklin. *Historical Collections of Louisiana, History of Louisiana from the Memoir of Dumonte de Montigny*. New York: Wiley and Putnam, 1852.

Guice, Julie. Marriages, Harrison County, Mississippi 1841–1899. Biloxi, MS: City of Biloxi & Harrison County Board of Supervisors, 1980.

Gutierrez, C. Paige. *The Cultural Legacy of Biloxi's Seafood Industry*. Biloxi, MS: City of Biloxi, 1984.

Higginbotham, Jay. *The Journal Sauvole*. Mobile, AL: Azalea City Printers, 1968.

———. *The Pascagoula Indians*. Mobile, AL: Colonial Books, 1967.

Hodge, Frederick Webb. *Handbook of American Indians*. Washington, D.C.: Smithsonian Institution, n.d.

Homes, Jack D.L. *Honor and Fidelity: The Louisiana Infantry Regiment and the Louisiana Militia Companies, 1766–1821*. Birmingham, AL: self-published, 1965.

Hulsey, F. Val Hulsey, PhD. *A Keesler Air Force Base, A History of the Site and Its Environs to 1941*. Biloxi, MS: Keesler Air Force Base, 1991.

Hutchins, Thomas. *An Historical Narrative and Topographical Description of Louisiana and West Florida Geographer to the United States 1784*. Introduction and Index by Joseph G. Tregle Jr. Gainesville, FL: University of Florida Press, 1968.

Inglis, G. Douglas, and Rodrigo Fernandez Carrion. *Colonial Family History in the Lower Mississippi Valley: A Workbook of Methods, Sources, and Bibliography*. Gulfport, MS: William Carey College, 1998.

Lowry, Robert, and William H. McCardle. *A History of Mississippi*. Jackson, MS: R.H. Henry & Company, 1891.

McLemore, Richard Aubrey. *A History of Mississippi*. Vol. 1. Jackson: University and College Press of Mississippi, 1973.

McWilliams, Richebourg Gaillard. *Fleur de Lys Calumet: Being the Penicaut Narrative of French Adventure in- Louisiana*. Tuscaloosa: University of Alabama Press, 1988.

———. *Pierre Lemoyne d'Iberville Gulf Journals*. Tuscaloosa: University of Alabama Press, 1981.

Moore, John Hebron. "South Mississippi in 1852: Some Selections from the Journal of Benjamin L.C. Wailes." *Journal of Mississippi History* 18, no. 1 (January 1956).

Le Page du Pratz, Antoine Simon. *The History of Louisiana*. Baton Rouge: Louisiana State University Press, 1975.

Phelps, Dawson, trans. and ed. "Narrative of the Hostilities Committed by the Natchez against the Concession of St. Catherine, October 21, November 4, 1722." *Journal of Mississippi History* (1953).

Purdy, Barbara A. *How to Do Archaeology the Right Way*. Gainesville: University Press of Florida, 1996.

Richmond, Stephanie C., and David Alfred Wheeler. *The Growth of the Biloxi Public School System*. Biloxi, MS: City of Biloxi, 1979.

Rowland, Dunbar, and Albert Godfrey Sanders. *Mississippi Provincial Archives French Domination*. Vol. 2, 1701–29. Jackson: Mississippi Department of Archives and History, 1927–1932.

———. *Mississippi Provincial Archives French Domination*. Vol. 3, 1704–43. Jackson: Mississippi Department of Archives and History, 1927–1932.

Rowland, Dunbar, Albert Godfrey Sanders and Patricia Kay Galloway. *Mississippi French Dominion*. Vol. 4, 1729–48, vol. 5, 1749–63. Baton Rouge: Louisiana State University Press, 1984.

Sheffield, David A., and Darnell Nicovich. *When Biloxi Was the Seafood Capital of the World*. Biloxi, MS: City of Biloxi, 1979.

Skelton, Zan. *The Biloxi Public Schools 1924–2001*. Biloxi, MS: Biloxi Public Schools, 2000.

Suarez, Julie. "The Biloxi Cemetery." Biloxi: Mississippi Coast Historical & Genealogical Society, 2002.

Sullivan, Charles L., and Murella Hebert Powell. *The Mississippi Gulf Coast, Portrait of a People*. Northridge, CA: Windsor Publications, Inc., 1985.

Swanton, John R. *Indians of the Southeastern United States.* Washington, D.C.: Bureau of American Ethnology, Smithsonian Institution, 1979.

Swanton, John R., and James Owen Dorsey. *A Dictionary of the Biloxi and Ofo Languages*. Washington, D.C.: Bureau of American Ethnology, Washington Government Printing Office, 1912.

Walthall, John A. *Prehistoric Indians of the Southeast, Archaeology of Alabama and the Middle South*. Tuscaloosa: University of Alabama Press, 1990.

NEWSPAPERS, CENSUS, DIRECTORIES AND OTHER RESOURCES

Bergeron, Kat. *Sun Herald.* "Rosetti's Café." 1996 and 2001.

Biloxi City Directories. Biloxi, MS: Biloxi Public Library, Historical & Genealogical Section, 1905–1958.

Biloxi Daily Herald. "Biloxi Loses Noted Citizen." December 24, 1926.

———. "Captain James Maycock." March 19, 1892.

———. "Captain James Maycock." March 26, 1892.

———. "Death Claims W.K.M. Dukate." March 29, 1916, 1.

———. "Elmer Funeral Largely Attended." December 27, 1926.

———. "Hundreds Gather to Honor Memory of W.K.M. Dukate." April 1, 1916.

.———."L. Lopez, Sr., Dead." September 26, 1903, 1.

———. "Mrs. F.W. Elmer Dead." January 23, 1905.

———. "Mrs. Julia Lopez Passes Away." July 1, 1918.

———. Reprint of General Holmes's letter printed in Charleston, SC *News and Courier.* "The Manor—Biloxi, Mississippi." September 25, 1912.

———. Sport Page. July 8, 1929.

———. Sport Page. July 9, 1929.

———. Sport Page. July 10, 1929

———. Sport Page. July 11, 1929

———. Twentieth-Century Coast Edition. Biloxi, MS: Biloxi Public Library, vertical files, Historical & Genealogical Section, 1900.

———. "A Venerable Lady Dies." April 11, 1900.

Biloxi Herald. "Aftermath of the Regatta." July 9, 1901, 4.

———. "Biloxi Regatta." August 12, 1929

———. "The Canning Industries of Biloxi." 1900.

———. "Closing." July 7, 1901, 1.

———. "Death Takes Wm. Gorenflo." February 17, 1932.

———. Fiftieth Anniversary Souvenir Golden Jubilee. Biloxi, MS: Biloxi Public Library, vertical files Historical & Genealogical Section, 1934.

———. "Impressive Service for Wm. Gorenflo, Sr." February 19, 1932.

———. "The Regatta." July 4, 1901, 1.

———. "Regatta." July 5, 1901, 1.

———. "Regatta." July 6, 1901.

Blessing of the Fleet, records and exhibits. Biloxi, MS: Maritime & Seafood Industry Museum, 2004.

Daily Picayune, September 5, 1892.
Fleur de Lis Society, records and exhibit. Biloxi, MS: Maritime & Seafood Industry Museum, 2004.
Guice, Julie. Lopez, Elmer & Company. Biloxi, MS: Biloxi Public Library, vertical files Historical & Genealogical Section, 1980.
Maritime & Seafood Industry Museum, *The Biloxi Schooner*. Seafood industry newsletter. Biloxi, MS: Biloxi Public Library, vertical files Historical & Genealogical Section, March 1982.
Stevens Collection. Papers of MS Coast Historian James Stevens. Biloxi, MS: Historical & Genealogical Section Biloxi Public Library, 1990.
Sun Herald. "I remember the night the Vancleave Special was born in Biloxi." 2007.
———. "Katrina Changes History." August 25, 2006.
Thompson, Ray Collection. "Know Your Coast." Biloxi, MS: Biloxi Public Library, vertical files, Historical & Genealogical Section, 1950s.
Tour Guide and History 1970–71. "Growth of the Seafood Industry." Biloxi, MS: vertical files, Biloxi Public Library, Historical & Genealogical Section, 1970–71.
Journal of Benjamin L. Wailes. Manuscript collection. Jackson: Mississippi Department of Archives & History, 1852.

Index

T

V

W

Y

About the Author

Edmond Boudreaux is a 1967 graduate of Notre Dame High School in Biloxi, Mississippi. He was an AT&T service technician from 1973 until 2010, when he retired. He is the administrative vice-president of the Mississippi Coast Historical and Genealogical Society; past president of the Gulf Coast Chapter of the Mississippi Archaeological Association; an acquisition committee member of the Ohr/O'Keefe Museum of Art; an advisory board member of the Maritime & Seafood Industry Museum; a member of the Community Advisory Committee for the Mississippi State Historical Museum; and a French colonial reenactor since 1992. He was awarded the Calvin Brown Award and the Mississippi Governor Commendation for archaeology in 1992. He was the City of Biloxi's Historian of the Year in 1993.